P9-CBC-843

THE PLANNER'S USE OF INFORMATION

ENVIRONMENTAL DESIGN SERIES

Series Editor: Richard P. Dober

Hemalata C. Dandekar/THE PLANNER'S USE OF INFORMATION: Techniques for Collection, Organization, and Communication
Joseph A. Koncelik/AGING AND THE PRODUCT ENVIRONMENT

Related Titles
Michael J. Bednar/BARRIER-FREE ENVIRONMENTS
Joseph A. Koncelik/DESIGNING THE OPEN NURSING HOME
Frank J. Smith and Randolph T. Hester/COMMUNITY GOAL SETTING

I define the term "environmental design" as an art larger than architecture, more comprehensive than planning, more sensitive than engineering. An art pragmatic, one that preempts traditional concerns. The practice of this art is intimately connected with man's ability to function, to bring visual order to his surroundings, to enhance and embellish the territory he occupies.

Richard P. Dober
Environmental Design

THE PLANNER'S USE OF INFORMATION

Techniques for Collection, Organization, and Communication

EDS/2

Edited by **Hemalata C. Dandekar**

The University of Michigan

Hutchinson Ross Publishing Company

Stroudsburg, Pennsylvania

To The University of Michigan urban planning students

Environmental Design Series, Volume 2
Library of Congress Catalog Card Number: 82-3119
ISBN: 0-87933-429-0

84 83 82 1 2 3 4 5
Manufactured in the United States of America.

Library of Congress Cataloging in Publication Data
Main entry under title:
The Planner's use of information.
(Environmental design series; v. 2)
Includes bibliographies and index.
Contents: Field methods of collecting information/ Hemalata C. Dandekar—Survey methods for planners/ Nancy I. Nishikawa—Information from secondary sources sources/ Allan Feldt—[etc.]
1. Regional planning—Addresses, essays, lectures. 2. City planning—Addresses, essays, lectures. 3. Policy sciences—Addresses, essays, lectures. 4. Communication in politics—Addresses, essays, lectures. I. Dandekar, Hemalata C. II. Series: Environmental design series (Hutchinson Ross Publishing Company); v. 2.
HT391.P543 307'.12 82-3119
ISBN 0-87933-429-0 AACR2

Illustrations by Yu-long Yang
Cartoons by Lisa Stout

Distributed worldwide by Van Nostrand Reinhold Company Inc.,
135 W. 50th Street, New York, NY 10020.

Contents

Series Editor's Foreword

The late Dennis O'Harrow, executive director of the American Society of Planning Officials, said it best, when describing the qualifications of successful planners: "At the top of the list I would put, approximately equal in importance, the ability to write and the ability to speak in public . . . unless you can do a better than passable job in both of these, you aren't going to function very well as a planner. . . . and I will say unequivocally that the graduates of city planning schools that I have seen recently have had this part of their training shamefully neglected" (*Planning*, August 1981, p. 25).

Hemalata Dandekar and her colleagues at The University of Michigan address such deficiencies systematically and constructively in their new book *The Planner's Use of Information*. They begin with a fundamental discussion of what kinds of information typically need to be communicated, how such information is gathered and evaluated, and the techniques for transferring to others the essential discoveries, evaluations, and ideas that are so basic to good planning.

Students and practitioners alike will find in this work valuable clues and explicit directions for improving the state of their art.

RICHARD P. DOBER

Preface

This book is about the various communications skills planners need to develop to become successful practitioners. The collecting, organizing, and delivery of information in efficient and convincing ways is critical to the discipline and is the essence of effective communications. A high level of verbal, written, graphic, and interpersonal skills is required in this endeavor. As planning schools in North American universities have begun to feel the need to teach these related aspects of planning, the lack of a convenient textbook for this purpose has become increasingly apparent. Currently, no single book pulls together the various strands borrowed from longer established fields or developed within the discipline and presents them in ways useful to both planning students and practitioners. Most reading lists for university courses dealing with some aspects of this subject are quite short and consist of scattered texts and journal articles that cover similar subject matter. The translation required to make them useful for planning is markedly absent.

Appropriately for a book on communications, this work is the product of a group effort. Although born out of a strongly felt need for material that would be helpful in the teaching of a course on communications in planning to urban planning students, its central themes and structure have evolved in the process of several meetings of the contributors as a collective. Many ideas occurred during the more informal discussions in smaller groups at various convivial places around the university.

Special credit must be given to Nancy Nishikawa for many hours of help in compiling the manuscript and to Rudolf B. Schmerl for assistance and encouragement beyond the call of duty of a responsible contributor. Lisa Stout contributed the cartoons, and Yu-long Yang drew the other graphics. Both worked rapidly and, we think, well. James C. Snyder has given critical encouragement from the inception of this project. Gerald R. Clark has made many useful suggestions about the content.

I am particularly grateful to the College of Architecture and Urban Planning of The University of Michigan, which provided financial support, space to work, and the use of equipment that greatly facilitated the preparation of this manuscript.

HEMALATA C. DANDEKAR

THE PLANNER'S USE OF INFORMATION

Introduction

On a warm Friday afternoon in May, Orville C. Lorch, mayor of Middlesville, Hiatonka, telephoned the city planning director to tell her that he had just met a delegation from the Middlesville Garden Club. The delegation had come to complain about the amount of trash regularly blowing around on Main Street and had demanded that something be done about it, adding several things about image and disgrace and more specific things about their role in city politics. "Kay," said the mayor helpfully, "why don't you look into getting some more whaddya-callems—not garbage cans, you know what I mean?"

"Some additional trash receptacles?" asked the planning director.

"Right—trash receptacles," said the mayor.

"People aren't using the ones we've got deployed in the central business district now," the planning director said.

"Well," the mayor answered, "if we put up some posters with catchy phrases, you know, kinda cute pictures and slogans, sort of, the good folks of Middlesville will pitch in. Get it? Pitch in. See what you can do, Kay." The mayor hung up, mopped his brow, and went home early.

SCENARIO 1: OLD METHOD

The city planning director called in Junior Planner, a recent graduate of the Department of Urban Planning of Hiatonka State University, and directed him to examine catalogs of trash containers, select a suitable receptacle, and come up with some "cute" ideas for posters by Monday. On Wednesday the planning director received a four-page report from Junior Planner, listing seven types of containers by cost, dimensions, materials, and manufacturers, and fifteen poster ideas. The planning director sent a copy to Mayor Lorch, noting that unless the mayor wished to contraindicate, she would present this information, along with her carefully considered recommendations, to the city planning board at its next Thursday evening meeting.

The planning director heard nothing further from the mayor, and both the information and her recommendations were referred by the planning board to a subcommittee created for the purpose. Those members of the planning board, doubting the wisdom of creating a subcommittee, did, however, succeed in having the board send a memo to the city comptroller, asking for an assessment of the affordable range of costs. Eventually the comptroller complied; the subcommittee submitted a six-page report to the planning board, agreeing with the planning director's original recommendation (although phrasing it rather differently); and the planning board sent a memo, with copies to Mayor Lorch and the executive committee and officers of the Middlesville Garden Club, to the comptroller recommending the purchase of a number of the desired trash receptacles.

Once the comptroller's request for authorization of an unscheduled capital outlay exceeding $500 had been favorably processed by the Finance Department, a matter not to be rushed in any community, the receptacles were ordered, delivered after a few delays, inspected, and installed just before the Labor Day weekend. Two weeks later, after the local high school's first home football game, the mayor received a delegation from the Middlesville Garden Club, complaining that the containers were ugly—"eyesores" was one of the gentler descriptions—that no one was using them, that the posters were being vulgarly defaced and used for obscene messages, and that, furthermore, they were blowing off, adding to the trash on Main Street. The mayor mopped his brow and went home early.

SCENARIO 2: NEW METHOD

The city planning director asked Mayor Lorch the names of the men and women who had called on him on behalf of the Middlesville Garden Club. The mayor's secretary was able to furnish a complete list. The planning director called in her Junior Planner, a recent graduate of the Department of Urban Planning of Hiatonka State University, and assigned him to spend some time that day and over the weekend on Main Street to get some impressions of the problem. Junior Planner was, furthermore, to contact the Garden Club members on Monday to see what they would suggest as solutions and to get some comparative data about downtown trash pickup from nearby cities of similar size.

By Wednesday, Junior Planner was ready to meet with the Middlesville Garden Club. In his discussions with them he agreed that the trash on Main Street was disgraceful. He told them that he had heard how impressed the mayor had been by the delegation's visit and was careful to mention each member by name. He showed them pictures in the catalogs he had brought with him of various kinds of trash containers, asked them to study the possibilities, keeping in mind costs and durability as well as aesthetics, and arranged to meet with them the following week to obtain their recommendations.

In the meantime, Junior Planner, following the planning director's instructions, continued to observe the situation on Main Street, taking photographs and talking to store owners and their customers. He learned that for two days after the regularly scheduled run of the garbage trucks on Tuesdays, the streets remained relatively free of trash. But by Friday the stores' dumpsters were full, and by Saturday they were overflowing. By Sunday the trash had spread along back alleys and sidewalks on to Main Street, and by Monday the garden club members' description of the scene was fully justified.

Junior Planner recommended to the planning director that the Board of Public Works reschedule garbage collection in the downtown district for Fridays or even Saturday mornings to minimize weekend buildup and scattering (as he phrased it). With the approval of the planning director, he explained this idea to the Middlesville Garden Club as one to be put to trial, after first demonstrating to them that their choice of trash receptacles, in the number they desired, would cost the city more than $15,000. He used his photographs and notes from his conversations on

Main Street as evidence that the notion was worth trying, and two weeks later the Board of Public Works, at a meeting attended by the planning director, Junior Planner, and the delegation from the garden club, agreed to change the trash collection schedule.

Two weeks later, the mayor, the planning director, and Junior Planner met with the Middlesville Garden Club. All agreed that the situation had been much improved—"much rosier," Mayor Lorch said, smiling broadly. The planning director and Junior Planner returned to their offices, and the Mayor, mopping his brow, went home early.

This story, with its alternate scenarios, was fun to make up when some of the authors of this book were wondering how to introduce what we intend to be a serious and useful discussion of effective communication in planning. But it would not have been much fun to experience it. Most of us will recognize in it some elements of our own experience in the practice of urban and regional planning. Planning schools have traditionally concentrated on teaching students a variety of substantive skills through courses either on theory or on various analytical tools. This orientation to teaching either concepts or methods and techniques is characteristic not only of schools that train researchers in a program leading to the doctorate but also of those providing the professional master's degree whose students become practitioners in the field.

Effective planning practice, however, requires the professional not just to integrate theory and analysis but to synchronize this with a planning process generally externally governed and outside his or her control. In this endeavor the planner's ability to communicate views, analyses, and ideas convincingly is crucial to success. Sophisticated conceptual models and complex analytic skills alone are insufficient. The planner has to extrapolate the essentials and communicate them convincingly to the client and interested constituents.

PLANNING PRACTICE

The professional planner is often required to formulate public policy based on information that is painfully inadequate for problem solving. Information about social and economic factors, pertinent to research on the problem at hand, comes in many forms. Sometimes it is straightforward, sometimes disguised or hidden, often outdated, and frequently not quite applicable to the problem.

Planning practice mandates that a planner obtain needed information as rigorously as possible and analyze data efficiently to arrive at policy conclusions within a defined time period (generally short), one often determined by political considerations. In addition to formulating policy, the planner designs, programs, budgets, administers, manages, and persuades using various verbal, graphic, writing, and interpersonal skills.

In the collection, analysis, and dissemination of information, the planner rarely works alone. Most of the time this person is part of a team of specialists who have diverse professional backgrounds and varied experience. To turn a group of independent individuals into a synergetic team of professionals that will produce required planning products on schedule is a task all practicing planners eventually must face. They must learn to work effectively as a member of such a group and on occasion to provide the leadership to make such teams work well.

Besides the interpersonal skills required to function effectively in a professional group, a planner has to learn to communicate with clients and constituent groups who will be affected by planning decisions. These consist of individuals who are informed and literate in varying degrees. Although the audience or client group changes its composition, particularly in the public domain, the planner has to maintain a continuous channel of communication to the groups that will be affected. Selecting appropriate media to do this with is important.

The mode of communication used in planning varies, depending upon the actors involved in the planning process; the language, which must be common to or at least understood by all the participants; the channels of communication, their structure, and points of contact; the scale and nature of the problem; and available finances. Knowing when to say what, deciding whether to write it, whether to send it, and to whom are skills essential to effective practice, ones usually acquired experientially, on the job.

The medium in which information is collected, analyzed, and presented in planning is a combination of written, mathematic, verbal, and graphic forms, supplemented by subtle, nonverbal, nongraphic, nonwritten interactions based on interpersonal skills and psychological and political acuity. The choice of the right questions to ask, the right techniques to focus on, and the most appropriate ways to communicate the findings effectively constitutes an entirely different set of professional skills that planners need to become effective and successful practitioners.

The fact that these skills, learned through thoughtful and aware practice, constitute the unwritten folklore in planning does not mean that they cannot be discussed, that no general guidelines can be proposed, that no observations can be offered that would help in their acquisition. This book has been designed to help planners learn some of the judgment, thinking, weighting, and intuition that go into making the "inspired" right choice of technique and method. It is designed to help both students and practitioners become more sensitive to what information is needed, how it is collected, and how it must be organized and disseminated. A planner's efforts at communicating information can be broadly categorized in three activities: collecting, organizing, and communicating information. This book has been organized accordingly.

The collecting, organizing, and communicating phases of a planning process are rarely linear but usually iterative. Planning decisions usually result from negotiations in which a problem is perceived, and needed information is collected from people, secondary sources, and on-site observations. This information is then analyzed in a group process, usually between specialists. The findings, and later the conclusions, are tested on a group loosely defined as a client or recipient. On the basis of feedback or another look at the context and additional information collection, another round of analysis occurs. During this iterative process there are flows of communications; sometimes one way, at other times an exchange between planners and clients.

Experience teaches a professional about relative weighting and the importance of considering various tacks in problem solving and communications. Initially when one has been given the training to execute particular aspects of a planning project, one tends to apply those skills and believe that this will make for success. As a result, planners tend to overwrite, overdraw, overtalk a project because they are relying on their best skill instead of picking up complementary skills in other areas that might be more crucial in effective overall communication. It is important to keep a clear perspective on what is significant in the decision-making process in order to assess: what counts, to figure out how to do it (what one must beware of and what to do in the way of checks and balances), and where and what to read and learn if the needed skills are lacking.

Before a planner decides which modes of presentation to use, he or she must assess what will make the most impact on the decision makers. For example, if public, verbal presentations are the major sales mechanism for a project, then the planning firm should hone its skills for effective verbal presentations. If mass media releases are crucial, the planner should know how to write them. It is of little use for a planning firm to spend most of its energy compiling an elaborate report with in-depth, sophisticated analyses and complex graphics if the people who will make decisions are not going to read it or to understand it. At the completion of a project, planners must assess what modes of communication were the most significant so as to better allocate staff resources for similar projects and contexts in the future. A review of the importance of effective interpersonal lobbying versus mass media publicity versus substantive, technical analyses can illuminate the potentially most fruitful expenditures of the planner's resources. This book highlights the need for this weighing of alternative modes of communicating. We hope that this book can serve to reduce pressure at moments of anxiety in the critical decision areas of the planning process and that it can provide some clarity about the organization and structure required to accomplish particular tasks in planning.

This book seeks to teach readers how to anticipate experience, how to supplement it, and how to amplify it. It anticipates experience by using hypothetical case studies of typical planning situations to illustrate the weighting process underlying the right choice in communication. It supplements experience by its broad coverage of a number of topics, including the uses of computers in planning, the appropriateness of simulation/gaming, consideration of small-group dynamics in work groups, and similar recent developments. The text provides overviews of such evolving areas and supplies an annotated bibliography for each topic. It amplifies experience by providing guidelines that can be consulted when a planning problem arises, an important presentation is nearing, a team of consultants is to be formed, or a report or graphic document is to be compiled. This is a practical book designed to be read in totality as well as kept handy for reference.

ORGANIZATION OF THE BOOK

The material covered in this book can be separated into two major categories: substantive areas emphasizing techniques of information gathering, organizing, and delivering, and processes, such as community participation and small group dynamics. The emphasis in each chapter is on problem definition and diagnosis, leading to selection of appropriate technique, method, or organization. The objective is to help planners choose the best way of setting about a particular professional task. The techniques described are illustrative, rather than all-embracing, and the descriptions indicate how careful choice among various communication techniques is critical.

The book is divided into three parts, addressing three kinds of questions:

1. What kinds of information does a planner require and where and how does he or she obtain it?
2. How does the planner organize this information?
3. In what ways can the planner communicate information and ideas?

The organization of this book into three parts with the chapters, and activities contained in each is illustrated in the accompanying figure.

To help unify the discussions in the three parts of this book, we have devised three typical planning situations at the neighborhood, city, and regional scale. These case studies are referred to in the book's chapters to provide contexts for the illustrations of what planners can do to make their communications more effective. The case studies are an integrating thread linking the chapters. Chapters begin with an introduction that discusses why and how the material covered in the text is important and relevant to the current and future practice of planning. They conclude with some topics for further consideration and an annotated bibliography. We hope that this book will convey the importance, even urgency, of becoming more effective communicators in the profession and also enable planning practitioners to develop some new skills, preparing them for new assignments and responsibilities.

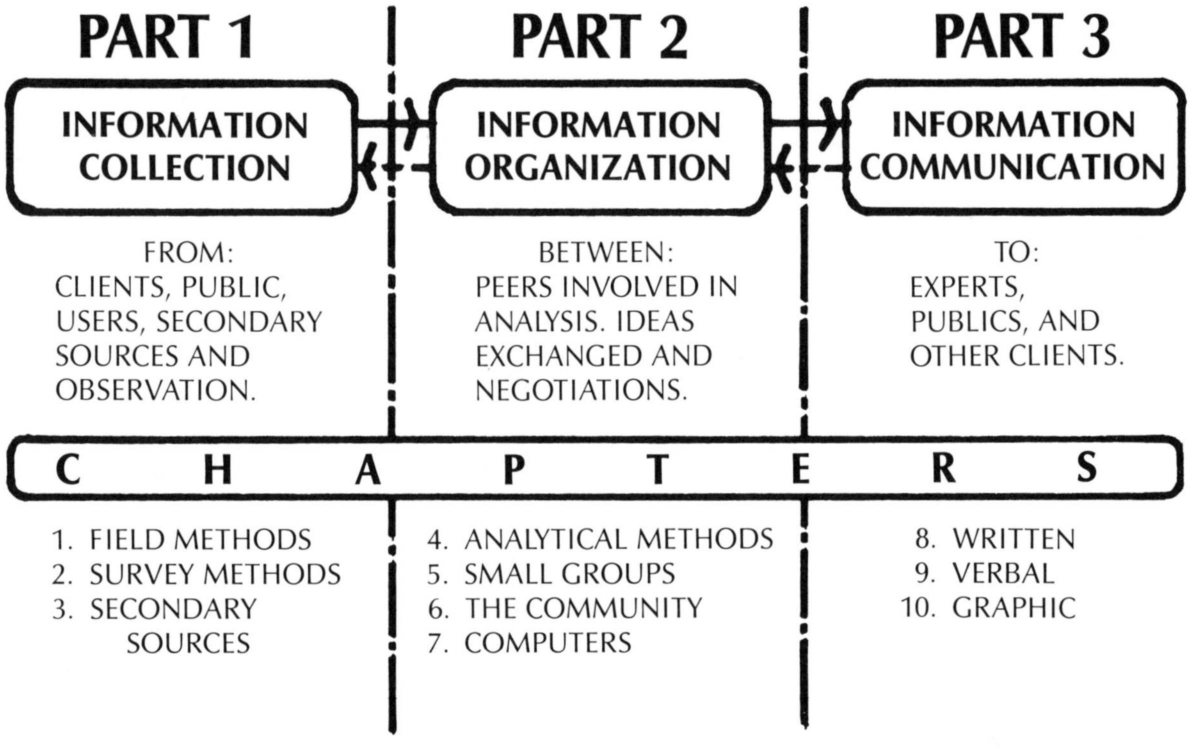

Organization of the book.

Planning Case Studies

The planning problems in the three hypothetical case studies presented here are representative of a range of problems (in issues as well as geographic scales) experienced by planning practitioners. They serve to illustrate various interactions where effective communication skills can be critical in affecting decision making. By referring to them we hope to present realistic contexts in which the techniques described in the chapters that follow can be successfully applied.

The first case study of Middlesville, with the story of Junior Planner, is referred to in several chapters of the book. The second and third cases are less elaborated. The second, of City Opportune, has been expanded in Chapter 6 on community participation; the third, about Hiatonka and East Victoria, sets the stage for the process described in Chapter 5 on working with small groups.

Unlike our story in the Introduction of Junior Planner in Middlesville, the problems described in these cases have no solutions, nor do we postulate a resourceful planning director or an astute junior planner. Insofar as we could avoid taking sides, we have done so. Heroes and villains are not our subject in this book; clear thinking and effective communication are.

NEIGHBORHOOD SCALE: URBAN DEVELOPMENT IN MIDDLESVILLE

A developer has bought several parcels of land in a transitional area on the fringes of a revitalizing central business district of Middlesville, a university town of 60,000 in the Midwest. Existing structures on the land parcels are eight older houses, rented by lower-income families or students, and two rundown commercial structures.

A private team of architects and planners, hired by the developer, has done a quick market assessment for the site development. They have recommended a high-density luxury apartment building as the most profitable development option for the site. Schematics for a fifty-unit, five-story-and-garage apartment building have been completed, and an application for a zoning change from R-1, C-1 (single-family residential with convenience commercial) to R-4 multifamily housing has been submitted to the city. The proposal is being reviewed by the city planning office.

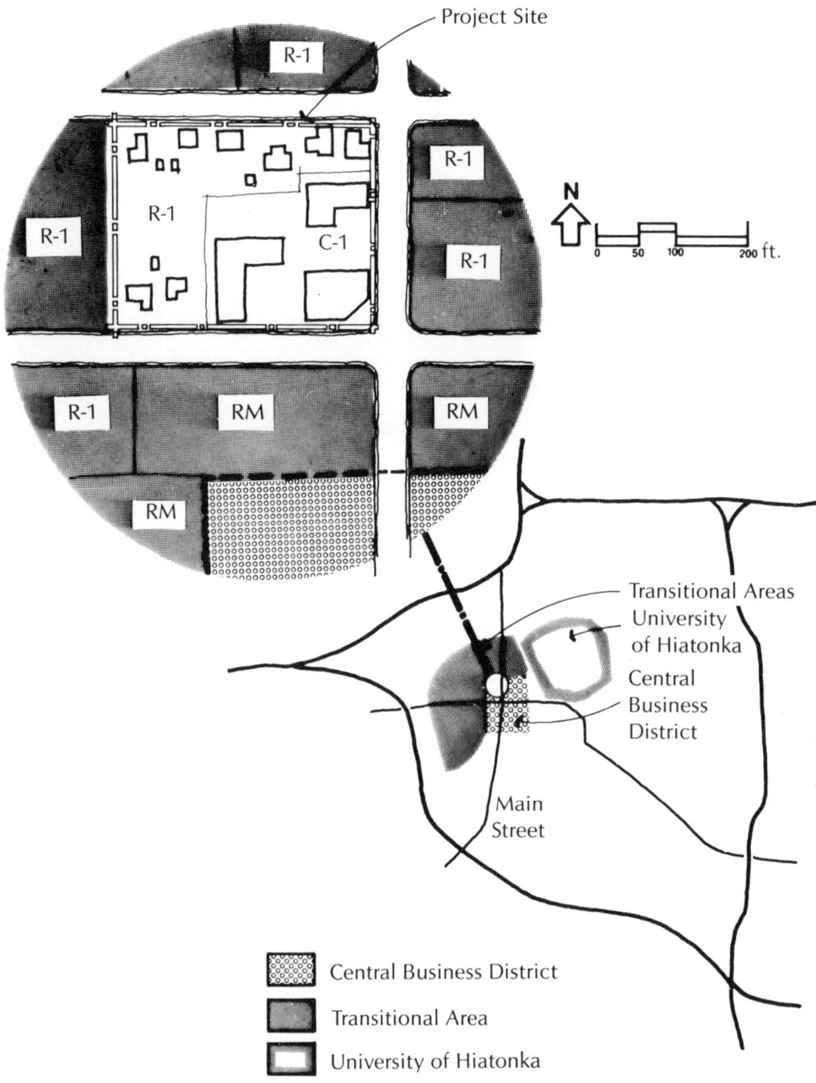

Development site in Middlesville.

According to established city procedure, the developer is scheduled to make a public presentation of his proposal before a regular planning commission meeting. As is customary, the agenda for the meeting has been announced in the local newspapers.

Meanwhile tenants living in the houses slated for demolition if the project is built have rallied the neighborhood with warnings of the gentrification process that will be started if this project is approved. On the other hand, local merchants in the central business district (CBD), delighted by the prospect of increased business and the influx of upper-income residents, have mobilized to support the developer's application for a zoning change and his request to reduce parking requirements for the project.

City officials are surprised by the large public turnout at the hearing. Reactions for and against the proposed apartments are vehemently expressed. This citizens' uproar signals a need to members of the planning staff to reexamine the project.

Since the planning department is developing a comprehensive revitalization plan for the CBD and fringe neighborhoods, the staff recommends that the developer's request be held in abeyance until their plan is completed in the next six months. At this time the controversial project can be evaluated in the context of the new master plan.

Upon this notice, the developer rushes back to his team with instructions to revise the proposal and create a more convincing development package with built-in strategies to "appease" the neighbors. The delay represents not just a financial cost to the developer but puts the whole site development in jeopardy.

The public presentation serves as a catalyst to mobilize the neighborhood residents against the proposed high-rise apartment. To strengthen their case, they team up with the local historic district commission that has been studying the area and propose designation of two of the homes slated for demolition as architecturally significant examples of early "carpenter" gothic structures from the 1890s.

Some representative and divergent points of view

Public official: "This is great! It will pump more money into Middlesville's CBD, and we'll be known as a city amenable to development and supportive of commercial enterprise."

Elderly renter: "I don't want to lose my home. I have no transportation to get to alternative housing suggested by the city."

Student activist: "I don't expect to live here, but I'm going to raise hell. The university should not let the city push the poor and students around."

Developer: "I own the property, and I have every right to develop it within the bounds of the law. The zoning change will benefit everyone. The project will bring more jobs and money to the city. The environment will be made safe for senior citizens."

City residents: "The project will cause parking problems near the CBD because the variance asks for a lower parking ratio. Furthermore, services in the area will deteriorate because of overload."

Journalist: "Is the mayor's wife working for the developer? How is it that the city has routinely been turning down requests for zoning changes and this one gets a hearing?"

CITY SCALE: TRANSPORTATION PLANS FOR CITY OPPORTUNE

The urban lobby in Washington, D.C., was able to pressure Congress to appropriate funding for studies of transportation needs in medium-sized cities (100,000 to 500,000 population) on the grounds that transportation systems will be a critical issue of the 1980s. On hearing this, the mayor of City Opportune phoned the head of the planning department and strongly urged him to submit a proposal to study their traffic congestion and to assess the need for mass transit.

City Opportune received a grant of $50,000 from the federal government for a preliminary transportation study. Highway department representatives contacted the planning department to bring to their attention a federal government program that would provide City Opportune with up to $6 million in subsidies if the transportation plan consisted predominantly of an expansion of the highway network. In addition to garnering additional and extensive federal funds, they argued, the city would be building upon the existing infrastructure and developing a tested and proved system. The automobile industry lobby had explained how an expanded highway network would benefit not just the auto industry but many other groups.

The city planners and their consultants had initially explored the feasibility of developing alternative forms of mass transport, including fixed rail, subways, bus systems, bicycle paths, and dial-a-ride. The highway lobby was persuasive, however, and a highway expansion plan was developed.

The city planners completed the study with the help of the transportation department and drafted a preliminary environmental impact statement. The plan was scheduled

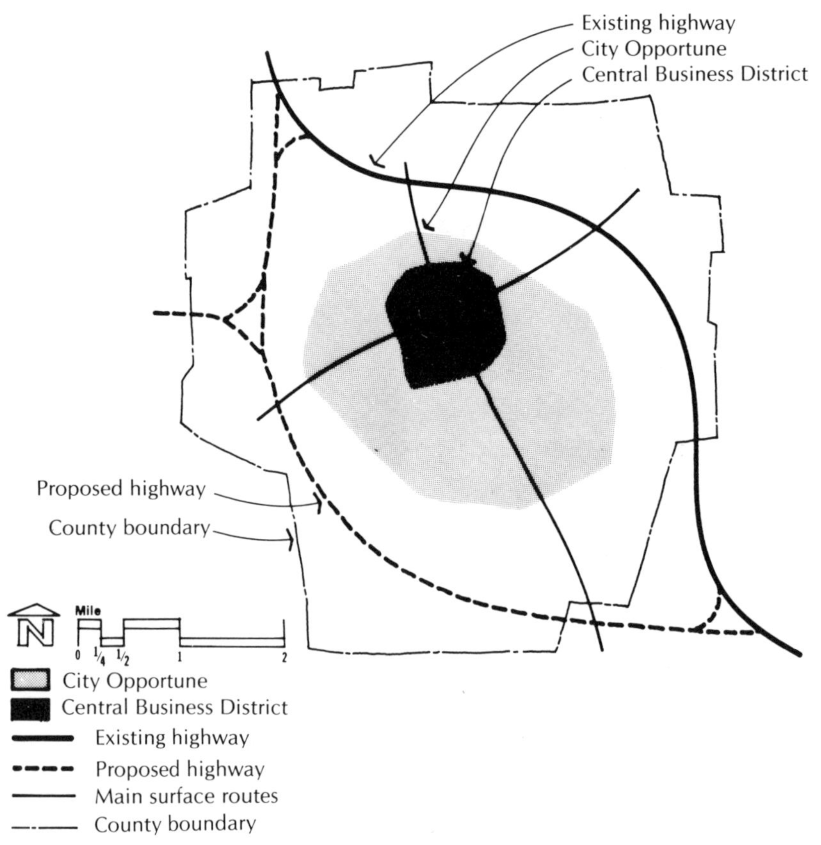

Proposed transportation plan for City Opportune.

to be presented before the city planning commission at a public hearing. A week before the hearing the major city newspaper printed an article on the transportation study contending, inaccurately, that the highway extension that the city advocated would cut a swath through one of the city's poorest neighborhoods and serve to segregate the "problem areas" from the downtown redevelopment project, currently underway, that it abutted. A public outcry ensued. The protests were centered not just on the highway issue but ranged from "no more taxes" to "we need more mass transit—now."

Some representative interests

Citizen: "I can't afford to keep a car so I have no use for highways. We need more public transportation."

Sierra Club: "Before anything is done, we want to get an assessment of the implications for air quality and wildlife in the only remaining wetlands in the area."

Antitax and antigovernment residents: "I'm against this!"

Industry sales representative: "We've got a deal you can't afford to turn down."

Rural area senator: "City Opportune already gets the largest share of federal monies for this state. Let's get some transportation money for our rural areas in order to attract industry out here and create jobs".

Low-income neighborhood residents: "They can take their highways and shove them through their neighborhoods."

REGIONAL SCALE: ENERGY PLANNING FOR HIATONKA AND EAST VICTORIA

The adjoining states of Hiatonka and East Victoria import over 90 percent of their energy requirements. Fuel industry lobbies are actively trying to get deregulation of fuel prices. Studies indicate that deregulation of liquid fuel oil will result in a doubling of the cost of fuel oil to citizens in Hiatonka and East Victoria and that the cost of natural gas will triple. Although availability of electricity will not be affected, its cost is expected to rise in proportion to fuel prices.

It is rumored on Capitol Hill that new federal oil pricing regulations could occur in the near future. The governors of Hiatonka and East Victoria, worried about the economic impact of impending price jumps, convince their state legislators to fund a study of alternative energy sources with a view to promoting self-reliance and use of local resources.

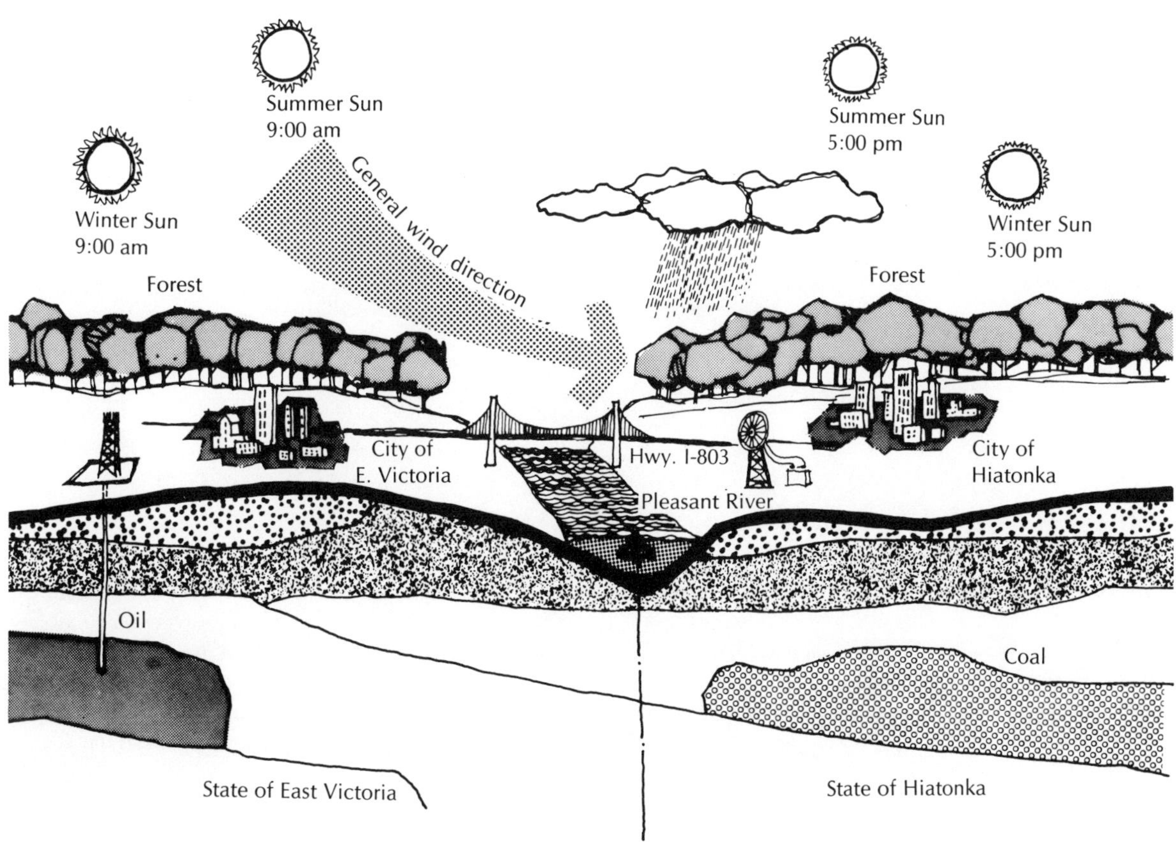

Energy resources in Hiatonka and East Victoria.

A private consulting firm, specializing in energy strategy, is hired to investigate use of wind, water, coal, solar, nuclear, and other energy sources. Consultants complete their study, circulate their report containing recommendations, and present their findings to a body of public officials. The energy department staffs in the respective governors' offices are expected to use the team's findings to formulate state policies in anticipation of changes in national energy strategy.

Some representative interests

Utility company: "We supply a major proportion of power in this region. We certainly want to be involved to make sure those 'advocates' don't make farfetched claims that cannot be implemented. We don't want to exclude coal as part of the alternatives or increased use of electricity."

Nuclear plant manufacturer: "We can provide plants on line that have served this nation for over a decade."

Environmental groups: "As soon as you put the bulldozer out, Mister Nuclear, I'm going to be lying in front of it."

Construction unions: "We need more jobs, and this project will employ 2,500 people for ten years."

Public service commission: "Let's talk about differential rate costs."

PART I

INFORMATION COLLECTION

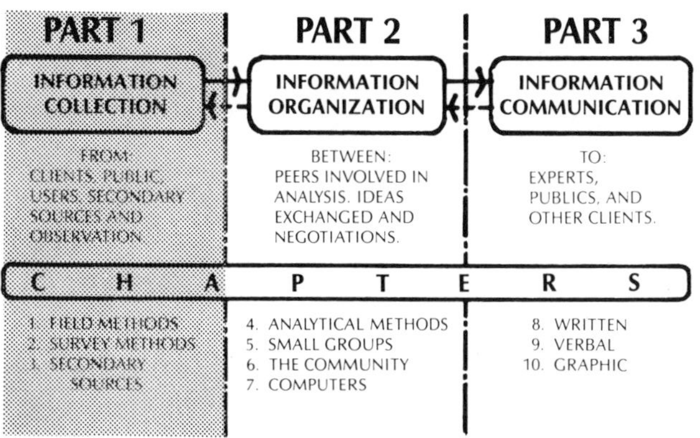

The articulation and elaboration of a perceived planning problem, as part of a search for appropriate policies and programs, require the collection of information: by observation of both human and animal behavior and the examination of physical plant, both natural and man-made; by administering questionnaires and carrying out and recording interviews; and by simple and complex projective techniques and creative examination of existing records. These different methods of gathering information, their strengths and limitations, are addressed in the three chapters in Part I. The questions addressed are: What kind of information does a planner need? How should he or she weigh and choose among the alternative forms it is available in? What methods should be used to acquire such information? Chapter 1 describes some of the direct, firsthand methods of gathering information in the field. Chapter 2 presents the different forms of information to be obtained from various types of surveys. Chapter 3 examines information available from secondary sources.

1

Field Methods of Collecting Information

Hemalata C. Dandekar

Encouraged by an enlightened city planning director, Junior Planner in scenario 2 of the Introduction in fact executed the basic elements of a site reconnaissance. Walking to the Central Cafe on downtown's Main Street, he took a firsthand look at the planning problem that had been presented to him: a problem communicated to him, as you will recall, by his director's account of the mayor's interpretation of the situation presented to him as observed by the membership of the garden club. Many problems often end up on a planner's desk in similarly circuitous ways. In view of this, getting some intuitive sense of the situation by collecting firsthand information can often be a good investment of time. In this chapter the various methods used to investigate conditions existing on the site, generically labeled *field methods,* are described, along with suggestions about which technique might be particularly effective for a specific situation. Included is a review of the steps involved in problem formulation in planning, the assumption being that the first concern in effective practice is the definition of the problem and an articulation of the kinds of information needed for further analyses.

That afternoon strolling down Main Street, camera in hand, Junior Planner photographed the pedestrian activity in the street, carefully remembering to include the level of trash both in the designated trash cans and blowing around in the streets. He paused to eat an ice cream cone at the street plaza on Liberty and Main streets and was eventually joined by numerous high school students who, he knew, had adopted this corner as their usual after-school hangout. This particular plaza was also heavily used during the day at lunch time by office workers. Junior Planner photographed and noted the trash level in that area during this interim period between the two sets of habitual users.

His cup of coffee at the Central Cafe, a place extremely popular with university students and staff, provided an opportunity both to converse with Joe the proprietor and his clients and to observe the comings and goings of people on that part of Main Street through the cafe's front window. Joe informed him that most of his suppliers made their largest deliveries of materials on Fridays in preparation for the weekend crowds. Joe therefore had to dispose of a number of bulky containers that evening or early Saturday, which filled his dumpsters to capacity and left them, at times, overflowing. He did not enjoy, he said a bit defensively, contributing to the trash littering the alley behind the cafe, but between the limited parking in the alley where the dumpsters were kept and the high cost of additional dumpsters, he felt he had little choice but to continue to let the trash pile up.

In eliciting this information, Junior Planner used some simple, basic methods to gather initial impressions, make a few observations, and collect some primary data. In gathering information in planning, practitioners select from the spectrum of techniques used in both the natural and social sciences. They choose from methods used in pure research for the development of basic theories and those that are most directed to immediate utility in applied research. Their choice of methods is subject first to a rough, almost intuitive, cost-benefit analysis of what is worth doing, evaluated on the basis of what information and insights it might yield that would be useful for policy formulation. The action-policy orientation of the discipline forces pragmatic choices in methods.

THE PROBLEM

Planning starts with a perceived problem and an articulated need for change. An information-gathering strategy evolves from this aimed toward a better identification and definition of the problem and an elaboration of what one wants to know about it and why. Given that scientific and systematic research for the practitioner is a compromise between the ideal and the possible in gaining an understanding of reality, some of the field methods described here, such as site reconnaissance, can be extremely helpful to gathering information rapidly and to assessing if the right questions are being asked about a situation that has been defined as a problem. Effective communication in planning, however, begins with the task of first accurately defining the problem.

In scenario 1 of the Introduction, Junior Planner learned, no doubt to his chagrin, that although planning, particularly in the public domain, may start with a perceived problem and an articulated need for change on the part of some subgroup of the public, an effective solution does not usually flow out of a direct response to the problem as it is presented. An analysis of the factors contributing to the condition as it is perceived and a reformulation of what constitutes a problem, in a more contextual frame, developed by casting a wider net on the issues and factors considered, is a necessary first step for the professional.

In the first scenario, Junior Planner played the traditionally accepted planner's role of a technical facilitator who accepted the mayor's and garden club's definition of the problem—accumulation of trash on Main Street—and their solution—more trash cans for its disposal—and proceeded to make a rational, systematic assessment of the most efficient and useful trash containers available. This course of action culminated in the selection of a container and its installation and led to embarrassing and wasteful consequences.

The premises on which problems are considered to be problems, by particular groups of people, require careful investigation to identify the specific nature and content of the complaint. Problems are, by definition, problems because of the observer's particular viewpoint and framework of reality, which will differ, often quite substantially, from others of a different social, economic, religious, or political persuasion or according to their sex, age, race, or nationality. The proverbial squeaky wheel attracts the most attention. A special interest group's formulation of a problem is predicated on the concept of reality shared by that group or held by the individual members.

Similarly the information that a researcher first thinks of gathering about a project or problem usually falls out of that person's internalized model of the problem and its potential solutions. The investigator's model helps define those characteristics that make a problem a problem and identifies areas about which more information is needed and why. The most important planning skill is to realize this is so and to be able to ask the right questions about a situation described as problematic. Once a problem is fully articulated and specifically defined, it contains the basic structure of the solution.

There are no universally right ways to do research, no one right way to do site visits or preliminary field investigations; there are only guidelines. Many different skills and methods can be brought to bear on an investigation, skills that will be enhanced through practice, experience, and learning by doing. It is wasted effort to learn the intricacies of a research technique—for example, sampling and significance levels of surveys—in isolation from a context. A method is best internalized by working on an actual and specific problem, and the process starts in learning to stop and ask oneself if the right questions are being addressed. The next step in a systematic inquiry is to test this speculative thinking, good idea, or theory. Its premises have to be confirmed, modified, or refuted by finding out, through empirical and/or other means, what appears to be reality. The major advantage of making a site reconnaissance trip and using various other field methods to collect information is that the researcher is exposed to a variety of stimuli, directly conveyed and received firsthand, the exposure to which can cause a reexamination of the premises on which the arguments and diagnosis of the problem are based.

In planning practice, a scientific and rational sequence of inquiry is usually not consciously followed. The political, time, and budget constraints and other pressures, including an atmosphere of crisis, are not conducive to such an explicitly organized approach. More often than not, the scientific method gives way to an attempt to complete something, almost anything, that the planner has the capability of producing in short order. Usually this is accomplished by drawing on skills and experience that the planning team is most comfortable with.

It is, however, a useful strategy to pause, in the normal headlong rush to meet deadlines, and jot down the steps in a more systematic sequence of inquiry:

1. What are you studying?
2. Why is it a problem?
3. On what concepts and models are you basing your approach?
4. On what theories or assumptions are these concepts based?
5. What do you think you will get out of this endeavor? What is the objective of this research?

Such a listing is useful in serving to clarify to the researcher what he or she is trying to achieve and why. A written list can be referred to when the complexities of the findings and the vast amounts of data start to muddy up the overall objective. A listing is also useful in the selection of methods that can help achieve the desired and specified ends.

SEQUENCE OF INQUIRY

Although it may not seem so at first glance, the methods of researching a planning problem do in fact follow most of the steps of inquiry developed in social science research. Most practitioners use procedures that, although at times sporadic and haphazard, draw on the methods established in more theoretical fields. Therefore it is important for planners to understand the discrete steps that constitute the so-called scientific method and to know how this relates to what is done in practice.

Selecting a Research Topic

The investigation can start from a variety of questions or problems that have one common characteristic: they are such that observation or experimentation can provide the needed information for some answers. Topics can be suggested by theoretical interest or dictated, as is more typical in planning, by practical concerns. The former are likely to involve a study of specific situations as illustrative examples of a general class of phenomena—for example, in the energy planning case study in Hiatonka and East Victoria, a laboratory test measuring the efficiency of two different energy sources. The latter are likely to involve study of a situation for the sake of more information about that particular situation—for example, in the transportation research for City Opportune, measurement of general traffic flows or the relative importance of alternative modes of transport in that particular city for populations in different socioeconomic strata.

Problem Formulation

Problems must be made concrete and specific. The topic chosen must be narrowed so the task is manageable in size and can be completed within a single study. This is a particularly difficult step for planners: to choose the right level of openness and closure in defining what elements of a system they will study, and to what detail. On the one hand, there is an urge to ask the seemingly peripheral but potentially interesting questions about a problem and, on the other, to hone in on what appears at first glance to be significant, in the interests of efficiency and under pressure to develop a finished product in the given budget and time frame.

Hypotheses Formulation

The role of hypotheses in research is to suggest explanations for certain facts and to guide in the investigation of others. Hypotheses may be developed from various sources.

A hypothesis may be based on a hunch or intuition, which may ultimately make an important contribution to the discipline. However, when a hypothesis has been tested in only one study, there are two limitations to its usefulness. First, there is no assurance that the relationship between two variables found in the given study will be found in other studies. Second, there is no clear connection with a larger body of knowledge. A hypothesis may arise from the findings of other studies, in which case it is freed to some extent from the first of these limitations. A hypothesis stemming from findings of other studies and also from a theory stated in more general terms is freed from both limitations.

In many areas of social relations, particularly in the field of planning, significant hypotheses do not exist. Much exploratory research and learning through trial and error must take place before such hypotheses can be formulated. Exploratory work of this nature is a necessary and essential step if progress is to be made in building up a theory of planning. Such theories are not of just academic interest; they are important to practitioners because they can help to expedite the problem-solving process involved in practice.

Relating Findings to Other Knowledge

Research can be a community enterprise in that each study can rest on earlier ones and provide a basis for future ones. There are two major ways of relating a given study to a larger body of knowledge: to plan the study so that it ties in with existing work at as many points as possible or to formulate the research problem in more abstract terms so that findings from the study can be related to findings from other studies. Research for planning problems can thus be done with two objectives in mind: to resolve the immediate dilemma, drawing on past experience, and to attempt to relate it to more generic problems for help in future problem solving.

Research Design

Research activities can be classified under four general groups, according to their major objectives.

The first are formulative or exploratory studies. These have the objective of formulating the problem for more precise investigation or for developing hypotheses. The scarcity of research in planning makes it inevitable that much of this research will be of a pioneering character. Certain methods are likely to be fruitful in the search for important variables and meaningful hypotheses:

1. A survey of the literature, particularly in related fields, for concepts and hypotheses that may serve as leads for further investigation in the planning field.
2. Collecting information from those experienced in the field by interviewing a selected sample of practitioners in the area, focusing the interview on what works and what are the change-producing agents. This approach is frequently taken in planning.
3. An analysis of insight-stimulating examples. The attitude and integrative powers of the investigator coupled with the techniques of analysis described in Chapter 4 are major factors in the success of this approach.

Second are descriptive studies. These use a wide range of techniques but are not as flexible as exploratory studies. The aim is to obtain complete and accurate information. As a result procedures used must be carefully planned, and many provisions must be made to guard against bias. The steps taken in a descriptive study are:

1. Formulating the objectives of the study.
2. Designing data collection methods so that they are as bias-free as possible.
3. Selecting the sample.
4. Collecting and checking the data.
5. Analyzing results. The analysis should be planned in detail before starting the actual work, including the statistics to be used and the coding method.

Third are studies to draw causal inference from experiments and fourth are studies to draw inferences from other completed studies. Information gathering toward these objectives is rarely attempted in professional planning so these are not discussed here.

This listing may serve as a useful guide to help clarify the specific nature and goals of the planner's search for relevant information in the practice of planning. Once the nature of the information needed has been systematically defined in this way, the planner can determine how to go about obtaining it. Chapter 2 describes the steps needed to construct a quick survey and the ways one might go about developing it. If the information might be obtained from secondary sources, Chapter 3 describes where and how to go about getting it.

Professional planners should be secure in the knowledge that the techniques of gathering information that they use in the field, in the form they use them, given time, budget, and personnel and skill constraints, are legitimate and properly founded on and drawing from a research tradition. The pragmatic approach that planners choose is justified by many of the reasons mentioned in this chapter. This does not, however, fundamentally affect the worth of the findings.

Organization

In planning, information that is gathered in the field is usually organized under the following nonexclusive and overlapping categories: spatial, social, economic, political, and historic. This categorization enables planners to understand a complex reality. But the planner's task is not to disaggregate in-the-field findings into these separate compartments but rather to integrate and to establish the relationships and connections between them. For example, the planner visiting the transitional area project under debate in Middlesville might record the age and condition of housing and its state of repair or disrepair and relate this information to the economic activity, racial mix, and historic evolution of the area and the politics behind the allocation of public funds in the city.

To collect the necessary data effectively, a number of information-gathering techniques—in the field and from other secondary sources—are required to establish some checks and balances. Conclusions based on findings from more than one source will have more credibility for all concerned: the researcher, the policy maker, and the affected population groups.

Data Collection

A planner's task devolves into three major categories: (1) collecting and synthesizing information about the problem at hand, (2) analyzing it to generate alternatives for action and to define and formulate a strategy for intervention, and (3) communicating these observations and findings in a variety of forms to different groups and constituencies.

These three categories of activity are not mutually exclusive or sequential but iterative and ongoing. When the developer in Middlesville listens to the reaction of the citizen groups to his development proposal at city hall, he is involved in communicating his plan of action and in collecting and synthesizing information about the problems standing in the way of its implementation. When he returns to his consultant planning team and urges them to react creatively to the opposition by generating a more acceptable alternative, and directs them to find other opinion leaders and mobilize them behind his proposal, he is involved in analysis and strategy. Thus field methods of information gathering are useful throughout the stages of a research endeavor.

SOURCES AND CHOICE OF FIELD METHODS

Field methods in planning are drawn from a number of disciplines traditionally categorized into those developed in the natural sciences, in areas such as animal behavior and plant ecology, and those developed in the social sciences, in areas such as cultural anthropology. This division between natural and social sciences is somewhat arbitrary and not particularly helpful when it comes to the specific selection of a technique that would be useful for planners. For instance, field studies of animal behavior, especially primate behavior, have more in common with approaches and problems of field studies of human social groups, such as tourists in Yellowstone, street gangs in Harlem, or sun worshippers on Venice Beach in southern California, than with anthropological comparisons of mythologies.

It is difficult to learn how to do field research from records and descriptions of research activities. These usually screen out the methodological problems of executing the research in the interests of clarity and brevity and also, to some extent, in an effort to convey a systematic, scientific, and therefore unquestionable set of discoveries. A neat sequence of events and observation is described, which includes only those steps that were instrumental in getting the researcher from point A to point B. Generally literature on the specific methods and problems of field studies is lacking. This expertise is traditionally acquired through apprenticeship and personal experience in a research endeavor. Certain books in fields like cultural anthropology attempt to fill this gap by addressing themselves to identifying problems one might encounter in the field through elaboration of a case study. These serve to sensitize the researcher to the range of problems that can be encountered, but since each field study is unique, they cannot help greatly in the actual decisions and actions required in the field.

The social sciences exhibit a concern with justification of a relevant topic of investigation. It is accepted that values and personal judgments, in addition to available information, play a great part in the selection of research topics and the types of data collected. The ethical and moral responsibilities of the researcher toward the subject group, the informants, and other researchers are issues touched on in most discussions of field methods. In the planning field, where one is potentially intervening in areas of interest to a variety of constituent groups, such careful scrutiny of values and criteria that are brought to bear on the investigation of a problem is essential.

The manner in which a planner begins to gather information in the field about a specific site and a problem is determined by various factors. A major consideration is the location of the site vis-à-vis the location of the planner and the scale and budget of the planned work. Obviously local, neighborhood projects call for a different mix and sequence of methods from those on a national or international scale. It is relatively easy for Junior Planner to take a walk down to Main Street on Friday afternoon and make repeated visits to the area, to observe and record over time, to identify potential informants who can provide more, different, and completely new perspectives on the problem, and to request and carry out in-depth interviews or participate in local activity to see how the system works. It is somewhat more difficult and more expensive for the energy commissioner or consulting energy planner of the newly formed bistate energy committee for Hiatonka and East Victoria to take a quick look at timber production patterns and yields in the two states and assess the impact of changes in the costs of alternative fuel sources on the energy consumption of people in various economic strata. It would also be more expensive to estimate citizen preferences for alternative transport means or assess if anything needs to change in the transportation pattern in City Opportune and make conjectures about how to do it. In both cases, although field visits and firsthand information may eventually be gathered, the first tack might well be to examine existing sources of information for the initial analysis of the problem.

At times when dealing with a small-scale planning problem, one may find in fact that there is little secondary source documentation and statistics that is disaggregated enough to be useful. In the interests of preserving the anonymity of respondents, censuses are notorious about suppressing information at a microblock or neighborhood scale. There is little recourse then but to do the necessary research or to make assumptions. In either case, firsthand, in-the-field investigation can be illuminating and informative.

While beginning to gather information on a rural community of perhaps over 2,000 people, you may discover with a shock that there is not even a rudimentary map of the settlement. In order to start a systematic survey you may have to laboriously map the location of all the housing units yourself and, in the process, might learn of much more complexity in the social relations, embedded in the spatial layout, than you had imagined in your original model of the community. Field investigations are valuable in this way. They serve not only to provide information where none exists but also to update it where it does. A town like Middlesville may have surprisingly little documentation in the way of statistics, maps, and qualitative descriptive material that is current enough to be useful.

Besides scale and location of project, other

interlinked elements in the decision to choose a method are time and budget. The critical factor in a project is the question of deadlines for completion of discrete components of the work and the affordable personnel to do it. The experience, strengths, and pool of skills of planning team members are important variables that must be considered before a flurry of information gathering begins. The important question in the cost-benefit weighting is what yields can be expected for a given expenditure of time and other available resources and the predicted usefulness of the findings in problem analysis or policy formulations.

SOME FIELD METHODS

Junior Planner, in his first attempt at getting a preliminary contextual and holistic understanding of the problem, used a combination of methods that could be loosely characterized in planning jargon as site reconnaissance (the walk or drive down Main Street recording, photographing, and noting), participant observation (eating an ice cream cone with the high school students and having coffee with the college crowd), field interviews (the chat with Joe about the state of garbage in his back alley), physical accretion measures (photographs of the level of trash in various areas, at different times of the day and over the week), and use of secondary sources (review and analysis of computer-generated information regarding the trash pickup schedule for Middlesville). In addition, he could have reinforced his case with an economic analysis (a comparison of cost of installing new garbage cans versus increasing personnel to allow for weekend garbage pickups) or quasi-experimental techniques (putting in some temporary garbage cans, perhaps borrowed from the parks and recreation department, to observe if increasing the number of trash cans would result in less trash in the streets).

In this endeavor to establish the facts of a particular situation or problem, the following field methods may be particularly useful to the practicing planner: site reconnaissance, windshield surveys, interviews, participant observation, and unobtrusive measures. The recording of information obtained in the field can be done in the form of handwritten notes, filling in of questionnaires, tape recordings of either observations the researcher makes in the field or of interviews, photographs, maps, sketches, annotations and overlays of existing graphics, and in higher-budget projects, recording images and sounds with more sophisticated equipment such as videopack and film.

Site Reconnaissance

Site reconnaissances are usually made to get an initial body of firsthand information, both qualitative and quantitative, about a problem. The opportunity to collect qualitative information is a particular strength of site reconnaissance techniques. The strength of the methods used is that they call for direct contact with people and the physical plant to be affected and thus provide the opportunity for anecdotal or firsthand observational insights.

The visit to Main Street allowed Junior Planner to analyze or at least question the presuppositions about relationships in the area that had been articulated by the mayor, the garden club, and the city planner, permitting him to develop his own understanding of the problem. Such direct, firsthand, experiential knowledge can be surprisingly powerful. Statements based on it have a persuasive quality far beyond their statistical representativeness. The information Junior Planner acquired was largely nonquantitative and statistically not significant, but it allowed him fresh insights into the relationships at play on Main Street and revealed different facets of the initially stated problem. This enabled him to translate and reformulate the problem.

A site reconnaissance gives the researcher a broader and more integrated picture of the terrain, vegetation, scale, and quality of the built environment and infrastructure, the mix of people, their races, ages, sexes, and an indication of their economic position from observations of their dress, age and condition of automobiles, and quality of housing. Most of all, a site visit provides an opportunity to experience the quality of life in the area. Does it feel safe and inviting, or alienating? Is it well maintained and well served with amenities or rundown and lacking in services?

Before the energy alternatives teams for Hiatonka and East Victoria go too far in their discussion and analysis of alternatives, they would be well advised to make a site visit to the areas under consideration. A project to construct a hydroelectric dam having an ancillary recreation and tourism industry will succeed only if both the physical and human resources (land and people, to put it more simply) will allow it. A firsthand investigation can give some sense of this.

Windshield Surveys

As the name suggests, windshield surveys are quick site surveys, often made in some vehicle so that the observer can cover a large area to be reconnoitered and record initial impressions. If the area is small, reconnoitering by foot or on a vehicle like a bicycle may provide a closer scrutiny and allow for interaction with users, assuming it is safe to so interact.

A windshield survey allows for the recording of the general ambience of an area. The survey is generally repeated at different times of the day, during various days of the week, in different seasons, and when there are various special occasions, such as the high school football game in Middlesville. The extent and content of the site surveys are determined and limited by the scale, location, time, budget, and personnel skills constraints.

Experienced planners begin to develop certain methods for particular contexts. For example, when visiting a new or even a familiar city, an initial drive through the area, early in the morning before business traffic starts, is an excellent preliminary survey technique. At this time, even in a busy city, human and vehicular traffic is at a minimum, and the observer can concentrate on the physical plant itself and its characteristics, as well as cover wide areas. Understanding the context and site conditions over time was useful for Junior Planner in Middlesville.

Windshield surveys can be used for various purposes. Often a first objective is to inventory, map, and record graphically by methods such as photographs and sketch overlays the current conditions on the site and to update the available information and ensure its accuracy. This initial inventory should preferably be made with some awareness of available sources of secondary information that can be updated or clarified in light of these site observations. Using information from secondary sources, augmented by firsthand field observations, at times can be the least expensive way of getting needed data. The synthesis of firsthand, visceral information and available secondary sources must be continuous and ongoing. New interconnections must be identified, and the problem and mechanisms at work must be reassessed so that new information builds upon and amplifies or corrects the old.

Ideally a windshield survey is made by two or three people, especially if a vehicle is required: one person to drive, another to navigate, observe, and comment, and a third to record. Recordings can be of various types such as oral descriptions or maps that are modified, sketched over, and added to. The right recording medium must be selected for the work at hand. All that Junior Planner needed on his visit to Main Street was a note pad and pencil, a camera, and some time and patience. The Middlesville developer's team visiting the project site or interviewing political representatives for their views on the proposed project might need to make better recordings and documentation of their activity. (See Figure 1.1.)

Field Interviews

Interviews can be an effective and useful technique for gathering information in the field. The types of interviews can cover a wide spectrum from very informal (a chat across the fence with a tenant in the threatened neighborhood in Middlesville) to a formal, structured administration of a standardized questionnaire. The informal interview can be quite conversational in tone. The researcher may have some ideas of what she thinks may be important areas of investigation and can query the respondent in a loose and exploratory manner. At the other end of the spectrum, a highly structured interview may consist of a list of carefully worded questions, in a carefully planned sequence, that have been pretested and modified to allow for comparisons of responses from a large number of people obtained by different investigators covering a wide geographic area. In between these extremes are various combinations of formal and informal interview techniques. Among them is the structured but open-ended questionnaire that allows for the collection of some standard information and includes open-ended questions that allow the researcher the freedom to pursue subjects and areas that surface as important during the conversation itself or from observations of the person's environment, dress, speech, and other considerations.

The essential ingredient in obtaining good information is the interviewer's ability to establish rapport with the person being interviewed. There are no easy rules about how this can be achieved. Maintaining a pleasant and open demeanor coupled with a genuine interest in both the respondent and the information he or she is communicating is a useful start. It is helpful if the interviewer's style is nonargumentative, supportive, sympathetic, and understanding. Contrary opinions can be elicited indirectly by such statements as, "There are people who say . . . ." and "How do you feel about that?" Dressing and grooming so as not to be too different from those being interviewed is also useful because it reduces the psychological distraction of being sized up by the subject. Aiming to be just slightly cleaner and better groomed than the respondents may be a good choice.

Practicing the interview by becoming familiar with the questions and thus avoiding stumbling over words or phrases facilitates a good interview, as does being prepared to interpret questions that the respondent does not understand. The interviewer can be prepared with a number of questions such as, "You mentioned . . . " and "Can you tell me more about . . .?" that can probe and stimulate the respondent to answer the more problematic questions. Such questions designed to probe must be as neutral as possible so as not to bias responses. If several field investigators are being trained to administer a questionnaire, some common probes might be suggested so that, if they do introduce a bias, the bias will at least be common across the whole survey.

An interviewer should not be afraid of silences, to sit quietly and wait for more information. A simple query such as "Anything else?" may evoke the desired response. There are direct and indirect ways to ask questions, and the interviewer, in the context of the respondent, must choose which approach to use. The interviewer can introduce a broad topic and allow the respondent to talk on any aspect of it he or she pleases or can question individuals in groups or alone, recognizing that answers are affected by the situation. Groups can inhibit as well as stimulate detailed responses.

The type of interviews to be made (structured or unstructured) is determined both by the nature of the information needed (from qualitative and anecdotal

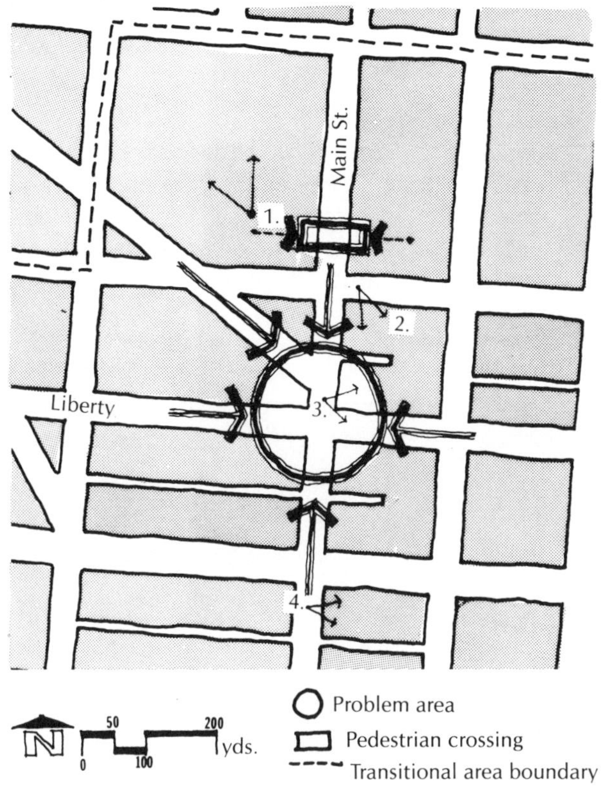

1. City Hall

2. Single detached family house

3. Religious institutions

4. Dilapidated commercial structure

Figure 1.1 Thumbnail sketch of a walk down Main Street.

to precise and measurable units of information) and by the size and characteristics of the population to be studied. The question of who should do the interviewing is also related. An appropriate choice of interviewers, in terms of age, sex, race, and class so as to be compatible with the population surveyed, deserves attention. A good choice can lead to richer and more complete information, a poor one to failure to get any that is useful.

Recording

The interviewer might choose to commit the conversation to memory if the interview is extremely informal, hypotheses searching, and generally fact-finding in nature. Usually, however, notes—from a mere jotting down of points raised to a complete documentation of what is said—are helpful. Even if the interview is unstructured, it is important to make an interview guide, which can be as simple as a checklist or inventory of subjects one wants to talk about. Some leading questions such as "What do you think about . . .?" can be jotted down on this guide to help the interviewer overcome long silences. In open-ended interviews the temptation to summarize, paraphrase, and correct bad grammar should be avoided in the interest of keeping the flavor of what is being said. Interviewers should try to record what is said exactly as given; marginal comments can elaborate on and interpret distinctive gestures or difficult-to-understand verbiage.

Interviews should be written up as soon as possible. They will be an amalgam of summaries and notes of general points, verbatim transcripts, and pieces of analysis or ideas. To help in bookkeeping, a face sheet, which is the front page of the interview, can be made listing factual data about the interview, such as name and/or number of persons interviewed, place, sex, age, and education. Critical data vary from project to project. The face sheet may be completed at the start of an interview or filled in later depending on what will set the best stage for the interview.

If the respondent is not self-conscious and does not object, a taped record of an interview, particularly an open-ended one, can be helpful. Freed of the need to make written notes, the interviewer can observe the subject and make side comments and notes of areas that require further investigation. However, too many hours of taped interviews, like too much statistical information, can be so intimidating in their volume that they are never used. Another danger is that the interviewer, relying on the tape recording, may not listen carefully. To avoid this, he or she should jot down at least key points. The interviewer must be selective about the method chosen for recording, selecting the form that is most appropriate.

The tape recorder should be a small, portable one that does not make distracting noises at critical junctures in the conversation. Interviewers should become familiar with its operation before taking it to the field. Observing the interviewer wrestling with a tape recorder or other equipment will not inspire a respondent's confidence. A tape recorder should never be used without the respondent's knowledge and permission.

Whom, when, and where to interview

Who is to be interviewed and when and where are important considerations. (Chapter 2 on surveys addresses whom to interview in various sampling designs. Chapter 6 elaborates some of the ways to identify opinion and community leaders.) Selecting people to interview in a preliminary site visit might be as random as stopping people in the street and starting a conversation. Restaurants, bars, bus stations, parks, and stores when business is slow can offer fertile ground in which to strike up such conversations. Finding a place conducive to conversation is important. Generally there will not be an optimum response from a passerby on a noisy, windy street corner on a cold day.

There are many ways to identify potential informants. The local papers can provide references to people who have participated in discussions of the issue. Representatives of religious groups, civic organizations, and organizational meetings on related subjects are others. Interviews with such people can be arranged on the telephone. Before making the call, the interviewer needs to develop a convincing explanation of what kind of information is needed from the person and a persuasive reason why the person should cooperate. Potential respondents might reject a request for interviews.

Usually no payment is required or allowed for information. At other times it can range from paying for drinks at the bar (or coffee at Joe's) to payment of an agreed-upon amount of money for a given amount of time. The ethical aspects of this must be left to each researcher to reconcile for each particular situation.

Information obtained from interviews, particularly that from informal conversations of random respondents, must be carefully tested in a larger framework. Individual defensiveness about a subject may give rise to inadequate or distorted estimates and information. Particular rationalizations may be widely shared in a culture, and if they are, they should be identified. Asking similar questions of a number of individuals, the same question of many individuals, and of the same individual over time, and then checking for consistency are some ways to validate responses.

Secondary sources can provide validation. For example, if some older residents in a low-income neighborhood of City Opportune complain of being discriminated against in the provision of adequate public transportation, a quick evaluation of city-wide budget allocations, disaggregated spatially, on public transport (easily accomplished with a computer, as discussed in Chapter 7) would validate or negate the substance of the statement. This would be an appropriate backup study to cross-check anecdotal information.

Comparing the perhaps glowing descriptions these citizens might give of transport facilities as they were in previous and better years to the picture that emerges after doing archival work in the library or the city's hall of records is another example of useful cross-checking. Anecdotal material can be a colorful and welcome foil to the more systematically collected secondary sources of information with which it must be validated but which often make for dull reading. Interviews are the means for obtaining anecdotes.

Participant Observation

A participant observation study is one in which a researcher becomes a member of a group or a participant in a social event under study to collect data about the other participants. There is a fundamental dilemma in the method: how secretive or open should the researcher be in observing and recording the activities of other participants? The dilemma lies in both methodological and ethical considerations and conflicts. A researcher who is too open about his or her objectives can affect the participants' actions and the dynamics of the group itself.

Gaining entry and acceptance to a group is the first hurdle a researcher must face. This can be an anxious and difficult period. Most investigators try to get introduced through a contact person who is accepted by the group. This generally helps to shorten the initial testing period the researcher is subjected to and provides at least one opportunity to establish contacts.

Assuming he is accepted in interacting with the group, the researcher acts as a change agent. He can provide, for the group he is studying, the same exposure to new ideas and culture as he gains from his interaction with them. His relationship with the group generally, and informants especially, is personal. Dependencies are established. Some studies require the establishment of sufficient friendship and rapport to allow for the needed confidences. Using this information can cause moral conflicts for the researcher. It is his ethical responsibility to be aware of the confidences that have been offered and to conduct his work so that the group studied does not feel exploited. If the group agrees to let future researchers experience the same opportunities he has been offered, he probably has won the group's trust.

Another danger is that in the process of participating fully, the researcher can begin to empathize so much with the group he is studying (termed "go native" in anthropology) that he loses some of the needed objectivity about his subject. In addition the biases and observations of the researcher can change with increasing familiarity with the group and the phenomenon being studied.

The researcher can fill a range of roles in a participant observation situation, varying from primarily observations of people's roles, relationships, and their interactions to primarily participation in the tasks the group is involved in and learning through this process. In the latter situation it is important for the researcher not to involve himself in mastering the task but in observing how and what others do with it. His involvement is a mechanism through which the group's condition is better understood.

A fieldworker tries to fill as many roles as possible—a master role that is constant and central to his being in a situation and as many subroles and relationships as possible. Different facets of a system are revealed in such encounters, yielding varied insights. The observer should establish a regular routine of observations, interviews, and recordings on a number of topics, some of primary importance and others secondary.

Most members of a group are not familiar with all facets of the group life so it is important to select and establish contacts with key informants who are familiar with different aspects. The researcher may find it helpful to compile a list of the key positions and match members of the group who might be willing to act as informants. Until the researcher can establish the reliability of each informant, he must be cautious in recording the data received from them. If there are language differences, the researcher should attempt to learn the language because interpreters tend to act as an extra interface. At the least the researcher should master enough of the language so that he can monitor an interpreter, if one is needed.

A participant observer needs to find a good vantage point from which to observe. Both physical location and social positioning are important. Junior Planner's selection of a window seat was an excellent choice for the limited study he was attempting. For more extensive work it is useful to locate oneself at the center of activity and to arrange living or working quarters to permit rapid and widespread observations. Covered outdoor work spaces and open porches are useful for such surveillance.

Researchers should maintain a regular routine of brief visits to particular places, people, or institutions in the primary area under study and take regular walks through the larger area, rigorously observing particular times and routes. They can make their own residence available for leisure activities with offers of food, games, music, or other activities commensurate with the group's interests. Researchers should avoid classification of observations in terms of their own cultural experience and try to understand the relationships and functions within the group under study.

An observer should try not to be too conspicuous in the way he takes brief field notes of his observations. Jotted notes supplemented by mental notes should be translated into a running log of field observations as soon as possible. Note writing may take as long as or longer than the time that was needed to make the observations or complete the interview in the first place. Both discipline and time are critical in maintaining useful records. Field notes should include maps, descriptions, comments, and any analytic themes that occur. Often in the process of writing notes, the observer sees integrating themes and concepts that begin to emerge.

Unobtrusive Measures

However well integrated an observer is in an area and in a group, he is still a foreign element with the potential to bias the information he is collecting. Webb et al. (1966) discuss methods such as measuring physical traces of erosion and accretion, referring to data periodically produced for other than scholarly purposes such as episodic and private records, and making simple observations and contrived observations using hidden hardware to avoid human error. They claim that questionnaires and interviews depend upon language abilities of the subjects and are costly and weaker in their capacity to study past behavior or change. They illustrate how some of the nonreactive measures can gain more accurate and cheaper information. If only a single method can be used for collecting data, a verbal report from a respondent is the choice. However, a good case can be made for considering other complementary measures and for using various methods to ensure some checks and balances.

An example of such measures in the category of simple observations is recording car license plates in various national parks, noting their state of origin to establish the distance draw of that particular recreational facility. Mechanical devices that measure vehicular or pedestrian activity are another fairly common method.

Measures of physical traces are a potentially rich source of information too often ignored by planners. These are measures of physical evidence not specifically produced for comparison and inference but that can be utilized by an imaginative researcher. In one of the most persuasive books that suggest possible sources suited for such unobtrusive measures, Webb et al. classify physical traces into two classes of measures: measures of erosion and measures of accretion. Traces of physical erosion cited as having been useful in some research were observations and measurements of selective wearing out of some material due to more extensive use than its counterparts. One example concerned vinyl tiles in areas fronting a particularly popular exhibit that wore out much more rapidly than those in other areas. Physical accretion measures are those that use the buildup of residues and patinas on surfaces and forms as indicators. The research measures are the amount and extent of deposits of material. Wall graffiti, the volume and subject material of posters on buses or subways, wear of park benches, and the density of fingerprints or nose prints on different store windows are other examples.

Planners often work with built form and space. Physical measures can provide information that contributes to an understanding of the activities and uses in that space that is useful in design. A well-known example is to provide paving stones and walkways on paths that pedestrians have worn as they take shortcuts on grassy public spaces.

Quasi-experimental techniques can also be imaginatively used in information gathering. A well-known application is to leave a locked car in different neighborhoods. The time lapse between its abandonment and vandalization can be used as one indication of the proclivity for theft and the lack of social control and social responsibility in the area.

Documents such as notes of meetings, journals, work books, and project files maintained on a continuing basis are a useful source of information. They can shed light on the observations made or actions taken in the time period covered in the record. For example, a researcher's field book in which she has merely noted the number of interviews completed every day and the weather conditions can be used to unearth a good deal of hidden information. The frequency counts of her observations of climate and reference to officially recorded weather data at an aggregate level can show if the weather was typical for that time of the year. Correlation tables (discussed in Chapter 4) can show if when she did fewer (or more) interviews, there was any perceptible and significant bias in the responses. When she did fewer interviews, did she lie on the beach for some of that time? If so, did that affect her observations? Careful scrutiny of her work book might prompt one to ask underlying questions and perhaps gain particularly useful information.

Attending a city planning meeting in Middlesville and following up with a review of the minutes of these meetings for the preceding couple of years could have given the developer some idea of the types of projects that could be expected to be controversial in that town. An analysis of the voting pattern of the planning board members could have provided him some understanding of the politics of the groups and of the coalition to be expected.

Another excellent technique used in planning has been the use of time-series photographs of particular public spaces, which are analyzed to give use patterns over the day or month and year. Films that have documented the use of this method are both enlightening and amusing. Although primarily used for the study of built form, this technique has great potential for collection of information on social dynamics and interactions between people.

CONCLUSION

Field methods are one set of information-gathering techniques that are particularly useful to planners. They can be used to update existing secondary source information, allow for a firsthand sense of the problem, and provide colorful and interesting insights.

To study a problem systematically, it first must be well articulated. The steps of inquiry needed for its further elaboration must then be outlined to serve as a guide. Various information-gathering techniques should be used so that the data collected can be cross-checked, the findings from one source confirming and reinforcing the findings from another. Information collected in the field makes a unique contribution in this endeavor by allowing for a holistic first cut at the problem. The internalization and firsthand identification thus evoked helps to break down presuppositions about relationships and can result in a reformulation of the problem itself. And besides all this, collecting information in the field is fun.

APPLICATIONS

1. Imagine you are a prospective developer visiting Middlesville for a long weekend. You are interested in developing housing or commercial property.
   a. What field methods of collecting information might you use to help you get acquainted with the town and enable you to select areas for further investigation?
   b. How would you develop a list of people you would like to interview to give you additional information or insights?
   c. What units of observation of neighborhoods might you make to categorize "declining" or "improving" areas?
2. Think of a planning issue that concerns you in your neighborhood—for example, lack of play equipment for young children in the public park.
   a. What questions might you ask, and of whom, to get a representative sample of residents' views on this issue?
   b. What observations might you make to substantiate that there is a need for such equipment?
3. You are the mayor of City Opportune and want to get some intuitive sense of who uses public transport in Opportune and for what reasons. How might you go about getting this information?
4. You are the planning director of City Opportune. You need some rough idea of current public transport usage in the city. The department budget does not include allocations of money

for a formal survey. Think of some unobtrusive measures that might provide you with the needed information.

BIBLIOGRAPHY

Babbie, Earl, *Survey Research Methods,* Wadsworth, Belmont, California, 1973.

Relevant to field methods are Chapter 9, "Data Collection II: Interviewing," and Chapter 2, "Science and Social Science."

Campbell, Donald T., and Julian C. Stanley, *Experimental and Quasi-Experimental Designs for Research,* Rand McNally, Chicago, 1966.

Gans, Herbert Julius, *The Urban Villagers: Group and Class in the Life of Italian-Americans,* Free Press, New York, 1962.

A classic case study illustrating participation-observation techniques.

Hanson, Norwood, *Perceptions and Discovery: An Introduction to Scientific Inquiry,* Freeman Cooper & Co., San Francisco, 1969, Chapter 9.

Pages 149-170 are useful to problem formulation and the conduct of inquiry.

Lofland, John, *Analyzing Social Settings: A Guide to Qualitative Observation and Analysis,* Wadsworth, Belmont, California, 1971.

Part 2 on the collection and management of qualitative materials is particularly useful; see especially Chapter 4, "Intensive Interviewing," and Chapter 5, "Participant Observations."

McCall, George J., and J. L. Simmons, eds., *Issues in Participant Observation: A Text and Reader,* Addison-Wesley Publishing Company, Reading, Massachusetts, 1969.

A collection of articles dealing with problems, issues, and solutions from the literature on participant observation. Particularly useful are Chapter 2, "Field Relations," including a piece by Raymond L. Gold, "Roles in Sociological Field Observations," and Chapter 3, "Data Collection, Recording and Retrieval."

Maccoby, Eleanor, and Nathan Maccoby, "The Interview: A Tool of Social Science," in *A Handbook of Social Psychology,* vol. 1, Gardner Lindzey, ed., Addison-Wesley, Cambridge, 1954.

Naroll, Raoul, and Ronald Cohen, eds., *A Handbook of Methods in Cultural Anthropology,* Columbia University Press, New York, 1973.

Useful reference for a wide range of material on methods primarily oriented toward theory testing and theory construction rather than description.

Selltiz, Claire, Marie Jahoda, Morton Deutsch, and Stewart Cook, *Research Methods in Social Relations,* Holt, Rinehart and Winston, New York, 1966.

Discusses on an introductory level the considerations that enter into the various steps of the research process in the study of social relations. The chapters discussing problem identification are particularly useful.

Simon, Julian L., *Basic Research Methods in the Social Sciences: The Art of Empirical Integration,* Random House, New York, 1969.

Particularly useful are the Introduction and Chapter 2, "The Language of Research."

Webb, Eugene, Donald T. Campbell, Richard D. Schwartz, and Lee Sechrest, *Unobtrusive Measures: Nonreactive Research in the Social Sciences,* Rand McNally College Publishing Company, Chicago, 1966.

Directs attention to social science research data obtained by methods other than the interview or questionnaire. The goal is not to replace the interview but to supplement and cross-validate it with measures that do not require the cooperation of a respondent and that do not themselves contaminate the response.

Whyte, William Foote, *Street Corner Society: The Social Structure of an Italian Slum,* 2nd ed., University of Chicago Press, Chicago, 1961.

A case study that illustrates the use of participant observation.

Williams, Thomas Rhys, *Field Methods in the Study of Culture,* Holt, Rinehart and Winston, New York, 1967.

Draws on the author's experiences as a cultural anthropologist in North Borneo. Detailed are: preparation for field work, selecting and moving into a site, the first month of study, setting up a routine of research, interviewing and setting up methods to verify data so obtained, choosing informants, and departure from the community and ethical responsibility.

2

Survey Methods for Planners

Nancy I. Nishikawa

The planning process begins with an investigation of the existing situation. Before the urban planner can develop and communicate a program of what should be, he or she must know what is. The more information available about people's actual needs and preferences, the better planners are able to satisfy them. However, few analyses of planning issues can be completely researched on the basis of available information. (Chapter 3 presents a description of information available from secondary sources.) Such data may not cover precisely the information a planner needs to know, or if it does, it may not be at the desired geographical scale (for example, measurements of housing characteristics are available but are aggregated at the level of census tracts that do not coincide with neighborhood boundaries) or omit critical variables (such as renter-owner breakdowns by race and income categories but not by sex or age groups).

Another problem planners frequently must contend with is obsolete data. Planners are called upon to formulate plans and policies that reflect constantly evolving circumstances, a task that calls for methods that allow planners to obtain information pertinent to the present client population. Yet dependence on data from other parties or agencies, such as the U.S. Census Bureau, means that current information is only available on a publication timetable beyond the local planner's control. The survey is widely recognized as a way of creating an area-specific data base, and sometimes even the hurriedly put-together survey can fill in a critical information gap. Nevertheless, because it requires large amounts of time, money, and skills, most planning offices do not carry out their own surveys. (An increasing number of local governments, however, have developed data assistance divisions, often as adjuncts of the planning department. Many of them serve as information coordinators for the various municipal departments, with potential benefits in efficiency and the ability to spread the costs of large projects such as surveys.)

To a large extent, planners always will be interpreting and working with data produced by the activities of others, primarily nonplanners. When surveys or other methods of gathering data firsthand have to be undertaken to obtain needed information about particular aspects of the problem at hand, the planning office may decide to contract for the services of a professional organization. Nevertheless the planner should be familiar with the components of the survey process to work with consultants fully.

Several techniques can be classified under the broader heading of survey methodology, including personal interviews, telephone interviews, self-administered questionnaires, and mail-in questionnaires. Each of these major survey tools is examined in this chapter. Another objective of this chapter is to look at the context in which a survey takes place—specifically to raise some of the considerations a planner should be aware of at critical decision-making points in the design and implementation of a survey. Attention is given to the concerns of the planners for whom the survey is more likely a short-term research project for troubleshooting on specific issues rather than an elaborate piece of social science research.

It is only appropriate to start off with an important caveat. Questionnaires and interviews and even observations by the so-called objective outsider are obtrusive measures. The appearance of the researcher and/or survey on the scene introduces an extraneous element in the normal course of social and physical interactions. Therefore measurements taken in the field reflect not only the situation a survey was designed to measure but also the particular interactions between the respondents and the survey instrument itself. Distortions caused by such effects can become a major source of error in any survey.

Another source of error results from the survey instrument's inability precisely to operationalize theoretically complex social issues. Development of a workable survey research instrument is carried out in three steps, each of which is a progression of the researcher's logic. The researcher must first define one or more concepts central to the study, for example, "quality of life." The second step is to choose indicators that allow an assessment of the phenomena. In one instance, planners may decide that housing characteristics comprise one element of a household's quality of life (the appropriateness of this determination would vary depending on cultural factors associated with the population being studied). The third step involves constructing questions designed to elicit responses, which can be translated or coded into some measurable form, such as frequencies, rankings, or percentages. With each succeeding step, the planner-researcher faces the problem of losing important information through error or by pursuing unproductive tangents.

Careful planning in the survey design and execution phases can mitigate but not eliminate the effects of error. Without a doubt there have been considerable advances in survey research, and planners should avail themselves of the various methodologies. Experts are able to calculate sampling error with great statistical precision, and planners are all too familiar with the impressive array of computer calculations based on survey data. However, this can also lead to a false sense of scientific validity.

Given the types of errors found in any single data-gathering process, planners must be diligent in cross-validating survey results against measurements and findings from other sources, including common-sense explanations. One of the factors that accounted for the success of Middlesville's Planning Department in resolving its downtown trash crisis was implementation of this kind of strategy. Junior Planner's information-gathering effort used several direct approaches: he consulted with the garden club, did a site reconnaissance, and talked with people working in the area. In this way, he was able to get a comprehensive picture of the actual conditions producing the trash.

SURVEY RESEARCH OBJECTIVES

The survey research process usually and, perhaps, properly begins with the planner asking questions rather than identifying data as such. The broader question, "What do I want to learn from the study?" precedes the more directed question, "What data am I looking for?" The answers thus formulated will constitute an initial statement of objectives. And by delineating the purpose and objectives of the project, a decision can be made regarding the appropriateness of a survey as the particular information-gathering device to be used.

There is no single correct formulation of the study problem. It is desirable to evolve specific objectives through discussions and decisions in which all members of the research and planning staff participate. Potential ambiguities should be clarified by a definition of the terms used in the survey framework. For example, in Middlesville, a survey "to assess the public's perception of the quality of life in the target area" may be instigated as part of the city's development of a comprehensive revitalization plan for the CBD and fringe neighborhoods. Yet "quality of life" can mean many things: the residents' satisfaction with physical surroundings, the social climate as a gauge of neighborliness, or economic well-being. Also whose quality of life is being measured is not clear. Do residents alone constitute the public, or should people who work in the area but live outside be included? Surveys typically include the adult population, but what about children whose living space is even more confined within the territorial boundaries of the neighborhood?

The group responsible for designing and conducting the survey should write down an initial statement that explains the specific purpose of the survey, the kinds of results expected, and the specific areas to be covered by the survey to meet these objectives. If the survey is part of a larger planning effort, the group should consider how the survey findings will be used as part of a later action program.

After the objectives of the survey have been decided upon, the procedures for it are laid out within the specified methodological guidelines and resource constraints. Any widening of the scope (for example, if the planning department then wanted to inquire about residents' reactions to more intensive or mixed land use in their neighborhood) should be permitted only if the main purposes of the survey are not adversely affected by spreading resources more thinly. When changes are deemed desirable, they should be accommodated as soon as possible in the survey development process so that new or revised questionnaire items also pass through the normal round of pretesting for bias. Mistakes and omissions are difficult to correct when fieldwork begins.

One means of avoiding later revisions is to encourage all interested parties to participate in the task of preparing a list of practical and/or theoretical questions to be covered in the survey schedule. As a planning aid, the survey is potentially a political tool. With expanding policymaking arenas, this tactic is advised as a way of preempting the formation of groups that later may exert a negative influence in opposition to the survey.

A written list will begin to sharpen the focus of the study and also define categories of data needed. Once a project is actually underway, there is a tendency to widen the survey's scope to obtain the greatest amount of information possible since the marginal cost of adding one more item to the questionnaire or interview schedule is far less than the cost of going back in the field to retrieve information missed the first time around. Nevertheless, trade-offs inevitably must be made between that longer, comprehensive survey and one that respondents perceive as an imposition on their time or of dubious validity because questions seem less relevant. These choices should reflect the different priorities of the planning body and the context in which the findings will be used.

Types of Survey Information

Whether the full survey is done in-house by planners or is supervised by them, normally it is accomplished at considerable expense. Therefore it is prudent to determine a good fit between the type of information a survey can provide and the type of information planners need for analysis. One typology distinguishing four categories of survey data is described below.

Profile data

Profile data refer to characteristics of the survey population. Usually it is not necessary for surveys to seek data on the whole range of possible personal characteristics. Relevant characteristics should be selected according to the problem focus. However, data are commonly collected on six characteristics: age, sex, marital status, race, family and household, income, and occupation. These variables serve to organize the information collected and allow the investigator to see patterns of relationships within the findings. Collecting such standard profile information can help to indicate when comparisons between survey results and other data sets, such as U.S. Census tabulations of various kinds, are appropriate. (A summary of Bureau of the Census surveys and publications can be found in Chapter 3.) Profile data are also an important part of many sampling procedures as this information is used to check the representativeness of the chosen sample vis-à-vis the study population as a whole.

Environmental data

Questions dealing with the environment identify certain facts regarding the circumstances in which the respondents live. Examples of this are data about the character of the immediate neighborhood, type of housing, or spatial characteristics such as the proximity of friends or relatives. This type of information can be used to map the boundaries of respondents' perceived action space—that area in which frequent social interactions occur. In a related fashion, environmental data can be used to suggest measures that will allow the planner to identify such concepts as neighborhood cohesion or neighborhood identity. Such information would be useful, for example, in identifying areas for Community Development Block Grant (CDBG) programs, as well as evaluating the outcome of specific projects.

Behavioral data

There is a gamut of relevant social behavior that can be successfully surveyed. In transportation planning, journey-to-work surveys are now widely used; this information is also available from recent national censuses. More extensive questions could also be included on a survey schedule—for example, distinguishing between use of public and private modes of transportation and different types of trips (commute, shopping, recreation travel, and so forth). Most surveys pertaining to behavior issues either ask about past actions, and therefore are limited by a time gap to after-the-fact analyses, or cover relatively major forms of habitual behavior for which generally reliable information can be obtained.

Psychological data

Data in this category cover a broad area of psychological information, including opinions, attitudes, awareness, motives, and expectations. From the planner's viewpoint, it is the area in which there are least likely to be data in secondary sources. Through surveys in the form of opinion polls, planners can get a glimpse of the qualitative dimension of behavior. Determining the presence or absence of attitudes and the reasons for holding them is another important survey objective: what kind of people approve or disapprove a particular position, and why? In turn, that determination is often influenced by people's level of information about the circumstances surrounding the issue. One cannot assume that issues and events are equally understood by everyone, and it is difficult to assess how people stand unless their understanding of the issue is also known. The motive concept stands not only for an individual's reasons for behavior but more generally for the forces impelling some action. Expectations represent a person's future orientation—that is, opinions and attitudes about what will happen or plans for future behavior.

When to Use Surveys

The information obtained from survey research is employed in a variety of settings. Survey data can be used effectively in situations requiring policy inputs, description and explanation, behavior modeling, and evaluation.

To obtain input into policy deliberations

Survey data can be seen as pieces of information from which the investigator hopes to develop a reasonably balanced and sensitive picture of a situation. The informed planner has a better idea of who in the community has what kinds of demands and how different demands relate to each other. The survey is an instrument through which policy makers can become aware of the views, actions, or expectations of a sampling of people who will be affected by their determinations. In this sense, planners who conduct survey projects can be seen as aiding a communication process among taxpayers, government officials, and consumers of public services. Nevertheless, the survey should be presented so that the planner's role as investigator is minimized, while attention is focused on the respondent whose cooperation is seen as furthering the community's good.

Survey data can be a source of valuable input into policy deliberations, but the data themselves are not directly translatable into coherent policy statements. Value judgments are still necessary to complete the analysis and to provide recommendations for subsequent action.

To describe and explain a phenomenon

Descriptive and explanatory surveys are undertaken when precise measurements about a given phenomenon are lacking. Data obtained through surveys expose a range of behavioral actions in the statistical sense. In particular, surveys are efficient tools for measuring the frequency with which an event or action occurs. Frequently in developing the central concepts of a planning study, and particularly when innovative use of theoretical concepts is being sought, key terms need to be clarified with quantitative information of the type provided by survey findings. For example, among other measures, the concept quality of life may be defined as a function of the number of rooms per person in a household and the proximity of a residence to open space areas.

When a survey is developed to test a hypothesis, the issue of rigorous survey design becomes especially critical. In this case, survey findings are used to subject a logically derived set of explanations or hypotheses to verification by empirical data. Hypotheses are supported or rejected by the data according to the rules of evidence firmly established in scientific tradition.

To develop models of people's behavior

The process of conceptualization is an attempt to make simple a world actually made up of many complex phenomena. The constructs or models that order an investigator's concepts are recognized as representations of limited aspects of the real world; however, they allow him or her to grasp difficult problems by studying the properties and components of key variables. Within this scaled-down setting, the planner-investigator can apply systematic procedures to determine whether a theory is consistent with actual observations.

To evaluate programs

When the same survey is repeated regularly, it is also possible to get an idea of changes over time. The information gathered can be used for before-and-after assessments of a program's goal attainment, cost-effectiveness, and other types of evaluation and monitoring. An important consideration for survey design is determining what aspect of the program is measured (and therefore on what basis success is to be evaluated). Appropriate design is further affected by the timing of the survey vis-à-vis the program's own stage of development, sponsor agency requirements, and funding needs.

When To Consider Alternative Methods of Information Collection

Investigators often use surveys to unearth sentiments generally unknown to planners working in their city hall offices rather than in the community. The exploratory aspect of survey research should not be ignored; however, overuse of survey methodology taxes the goodwill of the respondents. Rather than use surveys solely for exploration, when sufficient background knowledge is lacking to construct a well-conceived instrument, the investigator should ask if there are other ways of getting as much information, perhaps by carrying out field investigations, visiting deliberately chosen, representative areas of a designated place and talking with residents there. These techniques are discussed at length in Chapter 1.

Some categories of human behavior are less appropriate for measurement by surveys. First, surveys are not appropriate to obtain accurate information about a sequence of historical events. Instead of relying on the memories and impressions of many people whose involvements in those situations varied, archival research, checking through a newspaper's archives, or selective in-depth interviews with the key actors may be a more economical means of procuring information. Second, as discrete statistical measures, surveys are not good for tapping the flow of activities at the individual or group levels. An account that calls for measurement of continuous behavioral activities, such as the shopping patterns of downtown patrons, might be better obtained from a participant-observation study.

Factors Difficult to Control in Surveys

Working with people raises other issues that must be considered before embarking on a survey research project. To a large extent, the project's success depends on the cooperation of the respondents. Investigators can execute a survey more or less effectively depending on factors internal to the study such as the quality of the questionnaire, training of interviewers, and good administrative capability, but other dynamics are also at work in the population to be surveyed over which the research staff has minimal control. First, there is the level of interest about the survey's subject matter that is already present in the community. All other things being equal, a higher level of interest can be expected to generate a higher response rate, except very controversial issues or illegal activities where fear of indiscretion is present. Surveys that stretch out over a length of time and attendant publicity can increase interest in the survey. In such cases, the disparity in interest between earlier and later respondents could introduce error into the survey design.

Second, participation is affected by prior surveys that in all likelihood have no relation to the study at hand. This includes the number and frequency of surveys conducted previously and the impression they made. This condition, as well as lengthy surveying periods, generally are known as *contamination effects.*

A third factor is the relationship between the respondents and the investigators or auspices under which the survey is undertaken. Students and military personnel have a consistently higher response rate because they constitute something of a captive audience. Needless to say, in working with the general population, the perceived authority factor is considerably reduced, even nonexistent.

Finally, the respondent group's sense of security and privacy must be considered. Once confined to inner-city areas, today there is a generally high level of stress and fear of physical attack operating within a wider range of potential respondent groups, and many people are quickly put on guard when approached. To reduce the negative consequences of a wary public, one line of survey methodology has developed that is called *foot-in-the-door techniques* (Reingen and Kernan, 1978). Behind the actual techniques lies the idea that a survey is fundamentally a mechanism for communication. Therefore it can take advantage of related patterns of social interaction such as cooperation and people's desire to maintain a favorable self-image.

Investigators can alleviate some of the anticipated effects of suspicion by working under the aegis of a recognized and respected institution such as a nearby university. When it is desirable to avoid affiliation with a political body, the survey may be carried out under the name of a fictitious body organized specifically to conduct the survey. Then the ethics of conducting surveys require that the respondent's right to be fully informed about the purpose of the survey not be compromised, and it is always preferable to inform the local office of the Better Business Bureau of any surveying activity.

The question of respondents' rights deserves further attention. Planners who are conducting surveys as part of a federally sponsored project are likely to be affected by a law established to protect the privacy of persons who are the subjects of such projects. The pertinent statute is Public Law 93-348, Title II, Protection of Human Subjects of Biomedical and Behavioral Research, also known by its short title: the National Service Award Act of 1974. Specific guidelines are set up by the National Commission for the Protection of Human Subjects of Biomedical and Behavioral Research and have been published in the *Federal Register*. These regulations include federal clearance of questionnaires. In addition to the commission, all universities and colleges have established institutional review boards that review and screen the legal and ethical dimensions of projects covered by the law. The board can also help researchers to design surveys that meet the standards. Inquiries into the technicalities of survey participants' privacy and protection can be directed to the regional field office of the U.S. Department of Health and Human Services. Similar guidelines may affect projects with state government sponsorship.

THE ADMINISTRATIVE FACTOR

The administrative details of survey research are as important as theoretical aspects and should not be underestimated if a project is to be completed successfully. Determining financial and time budgets and then coordinating these with needed materials and personnel are far from being mundane. Whether the survey project is large or small, inadequate resources and ill-planned execution can throw a carefully designed survey out of kilter. High priority is readily given to the scientific aspects of surveying such as sampling and measurement, yet sloppy organization can introduce unnecessary error into the study just as easily.

The administrative plan should identify all components of the project that must be acted upon. Individual tasks can then be delegated along with instructions about how they should be accomplished. The survey organizers should have a clear idea of how the various tasks are related to each other. Organizational information should not be confined to a select few; all team members like to be aware of what is going on: how one person's tasks fit in with those of another, how much leeway is allowed for decision making at one rank (and when decisions should be referred upward), and how individual performance contributes to the successful completion of the entire project.

Planning the administrative end of a survey calls for careful consideration of resource expenditures. Miscalculation of funds to pay interviewers could mean that some portions of the sample must be dropped, thus leaving only responses from more readily accessible persons. This bias may be avoided by holding to the original sampling distribution but sampling a smaller number of people. Misjudgment about available staff resources could result in fewer callbacks than planned for and probably a less-than-desired response rate. In both such instances, the quality of data obtained would reflect regrettable decisions about resource allocation that impede the intended research design.

Travel and personnel costs usually account for the bulk of the sample survey budget. The project will likely need administrative staff, clerical staff, field staff, in-house professional staff (such as a statistician, an economist, and a health specialist), and outside consultants.

Estimates of travel costs and living expenses in the field should cover the time spent for sampling, testing the questionnaire, interviewers' training, and actual fieldwork. Interview costs are usually the largest item in the budget. However, a study made by the National Opinion Research Center found that, on average, only one-third of an interviewer's time was directly attributable to interviewing; the other two-thirds was spent on less critical tasks such as travel, research, editing questionnaires, and other routine clerical tasks (Sudman, 1967). Large survey organizations have budget allocations of $50 to $120 per respondent depending on the geographical scope of the survey.

The administrative plan should state the order in which tasks must be accomplished and the amount of time necessary for each. Like cost budgets, time budgets are often underestimated. Allocating contingency time beyond the normal time budget can mitigate the later occurrence of unforeseen obstacles and delays. Another way of building in contingency time is to work with core procedures while allowing the option of adjusting different tasks when expected outcomes do not result.

The project's budget, staff, and time schedule are interdependent. Laying out neat charts and timetables is a skill most planners have learned to carry out with dispatch and even flair; however, the research and planning staff should not hesitate to evaluate critically their available resources in the context of fulfilling the original objectives. If there is any doubt about successfully completing the survey or an inability to cope with emergencies is suspected, the option not to proceed should be considered early in the project effort. Short of canceling the entire project, however, it can often be reduced in scope better to match existing funds and personnel resources. Alternatively, when relatively small amounts of data are required, it may be possible to buy time in an omnibus survey conducted by a commercial or university-based survey organization, also referred to as *piggybacking*. Indeed the planner's relationship to a survey effort is frequently supervisory.

SURVEY DESIGN

In the project design stage, the specific procedures and methods to be used are established. These include selection of survey design, population to be sampled and sample size, method of administering the questionnaire, and construction of the questionnaire. Planners have a considerable number of alternatives from which to put together the final survey project. The major options available are briefly discussed here to serve as a starting point for relative comparisons between choices. For more detailed information, the reader is directed to the several excellent books listed in the bibliography of this chapter.

It is easy for novices to get caught up in the pros and cons of various techniques and strategies. There is an abundance of good advice, guidelines, and checklists (which should be taken seriously).

Nevertheless, one of the best ways to learn how to conduct effective surveys is to go out and do the work. With successive iterations, researchers will be able to devise a survey methodology that fits the requirements of the working environment and capitalizes on the skills and talents of the survey team.

Basic Types of Survey Designs

There are many ways of classifying the different types of survey designs but for the purposes of introduction or review, a useful typology is the distinction between the single cross-sectional survey and the longitudinal survey.

Cross-sectional surveys

Unweighted cross-section The one-time unweighted survey is probably the design most familiar to planners. It produces a "snapshot" for measuring the characteristics of a population at a given point in time. For example, as part of the CBD revitalization plan in Middlesville, the local planners may want to gauge public opinion concerning the attractiveness of the CBD as a shopping district. By polling a sample of persons representing all elements of the population in which they are interested, the planners will be able to assess public sentiments. (See Figure 2.1.)

Weighted cross-section Another version uses weights deliberately to oversample certain subgroups of the population. Usually this group has special significance for the survey but is known to constitute a minority within the total population. In such cases, an unweighted cross-sectional survey would not yield enough cases to complete a meaningful analysis. Thus a planning department interested in developing facilities for the annual ethnic fairs as part of the CBD plan would want to survey the subpopulation that would use these facilities most extensively. One way of reaching a large number of fair participants is to double or triple the sampling rates in census blocks having high concentrations of people in ethnic groups that sponsor fairs. (See Figure 2.2.) This assumes that people living in these areas have a greater interest in and tendency to use the facilities. (If some groups are oversampled, it is important to readjust those cases to their proper contribution of the total sample for data analysis.)

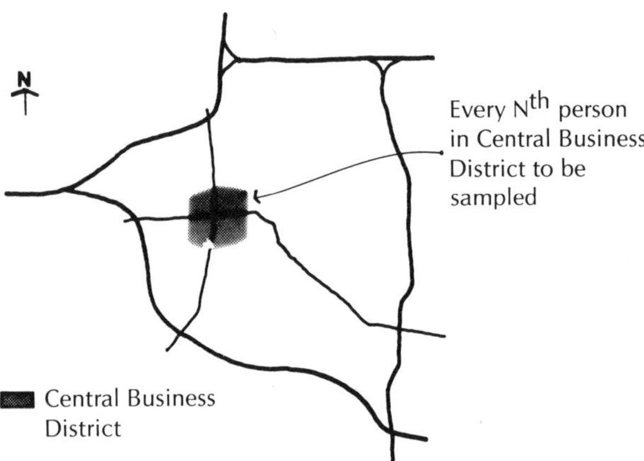

Figure 2.1 Unweighted cross-section.

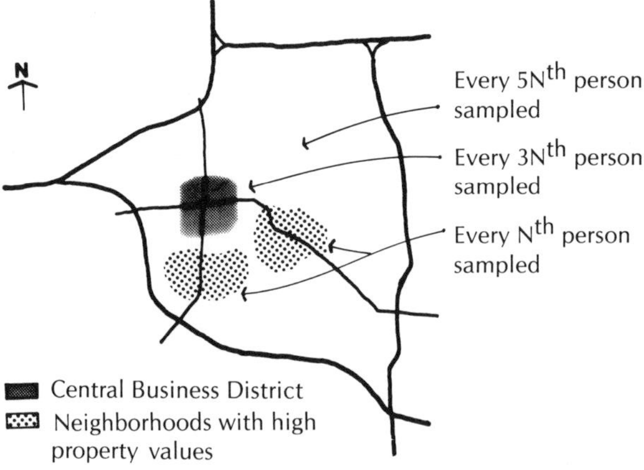

Figure 2.2 Weighted cross-section.

Contrasting samples Sometimes it is useful to draw samples from groups that are already known to show substantial differences with regard to an important variable in the study. For example, if the departure of industries from the CBD is a serious problem, planners may be interested in interviewing the employees of firms that have moved to a new location. The purpose of such a survey would be to discover differences in attitudes and characteristics of employees who have remained in the original locale and stopped working for the firm and employees who relocated and continued working for the firm. (See Figure 2.3.) Significant differences between the two groups could aid in population projections and revision of the local labor profile. These kinds of data are vital for planning regarding school size and location and recreation and shopping facilities, as well as for skills-retraining and other social services programs.

Longitudinal surveys

In longitudinal surveys the objective is to study the degree and direction of change in a situation rather than its static state. It is therefore necessary to obtain measurements at more than one time for comparison. The two types of studies that commonly take successive cross-sections of the same population are before-and-after and trend analysis designs.

Before-and-after study This design is used to measure the effect of some stimulus on a target population. Accurate assessment of the effect—whether there has been improvement, decline, or no change—requires measurement before and after the event has taken place. (See Figure 2.4.) The difficulty usually lies in obtaining the "before" or "baseline" data because researchers often begin their work after the stimulus. Nevertheless, with some ingenuity, it is possible to find ways of estimating or reconstructing the before situation, such as through respondents' recall of that information, other data sources, and/or related survey material.

Trend analysis The study of trends depends on more extensive monitoring, and measurements cover a longer period of time than just before and after a particular event. Periodic collections of data on attitudes toward transportation systems, education, housing, or revision of the zoning ordinance could serve as continuing social inputs in the planning process. The availability of computers has increased the attractiveness of extensive data bases by providing easy storage, access, and manipulation capability.

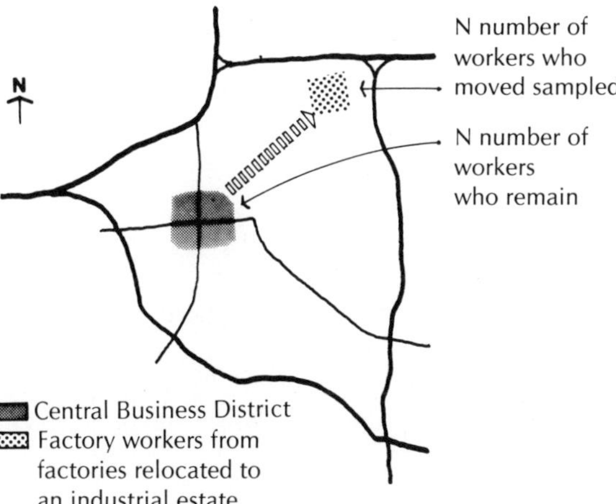

Figure 2.3 Contrasting sample.

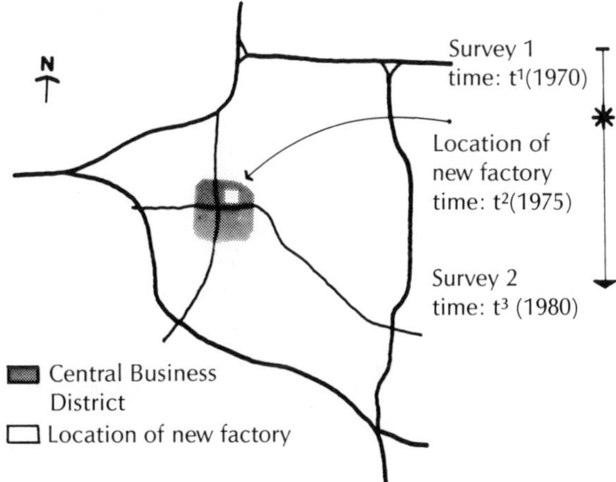

Figure 2.4 Before-and-after study.

One variation of trend analysis calls for the data to be collected predominantly from the same sample of respondents, called a *panel.* (There is usually some mechanism through which individuals can enter or leave the study. For example, if the panel is based on a sampling of addresses, a family that moves to another location is out of the study and the new occupants take their place as respondents.) The advantage of repetitive contact with respondents is that it allows the researcher to build up comprehensive data sets on individuals that almost amount to case histories.

Selecting an Appropriate Survey Design

Given the availability of several design options, the planner is left with the decision of which design to select. When he or she is interested in a one-time description, some form of the cross-sectional survey is most appropriate. Choosing the specific type would be influenced by the population relevant to his or her interest and should be coordinated with selection of the sample.

On the other hand, if the objective is to examine a dynamic process— one involving change over time or the identification of behavior and attitudinal patterns—then the longitudinal survey is more suitable. Such a study requires years to complete and usually needs the support system of a research program to provide continuity between actual surveys and to help meet large expenses.

The attributes of the simple cross-sectional survey and the advantages of the longitudinal survey can be combined by locating a prior survey that collected data relevant to the research subject. In such a case, the updated information provided by a new cross-sectional survey can form a basis for measuring trends. The planner using this methodology should be careful that the sample design and questionnaire items of the new survey are comparable with the older survey. Awareness of the limitations in the methodologies used by previous surveys is crucial, especially of the extent to which those findings warrant conclusions about any population besides the sample group.

SELECTING THE POPULATION TO BE SURVEYED

Probability Sampling

The findings of a sample survey accurately relate only to the population from which the sample is selected. Therefore it is critical to arrive at a clear definition of the *target population,* the body about which conclusions will be drawn. Of necessity, then, a sample cannot be chosen until the whole, or the universe that it represents, is decided upon.

Determination of the target population is made up of two elements. The first is specifying the spatial area to be covered by the study, such as a city or a neighborhood. Alternatively this may be a physical unit such as a factory, school, or dwelling. The second is selecting the actual person to be interviewed. This selection depends upon the nature of the information desired and practical considerations of convenience and access. When accurate information about a person can come only from that person, as is often the case in attitude surveys, it is essential to have a predetermined plan for randomly selecting persons within the household. On the other hand, if the survey deals mainly with information about the household and its occupants that any adult can provide, careful prior selection is less important.

References have already been made to the sample survey. Unlike the census type of data collection, which counts every member of a given population, a sample takes only a segment or portion of the population for the purpose of making estimated assertions about the nature of the original population. In particular, a good scientific design is based on the theory of *probability sampling* in which there is a known probability or chance that any individual person will be included in the sample. Ability to calculate the

representativeness of a sample is, conversely, the rationale for making generalized conclusions about the universe from the sample findings. Probability theory depends on mathematical calculations to determine how many respondents are needed to achieve a certain level of accuracy in the survey results. The investigator can also estimate a mathematically derived probability of error that would result from interviewing a sample of people instead of locating and interviewing each person in the universe. When properly executed, sampling can enhance the accuracy of data gathering even more than the full count census. From a practical standpoint, it is apparent that handling smaller numbers permits the collection of more detailed information. In many cases, much more can be learned from an interview of thirty minutes than six meetings of five minutes each. Furthermore because smaller numbers are handled throughout the survey, greater attention can be paid at all stages, from data gathering through processing and analysis. Additionally the costs of the investigation in time, staff, and inconvenience to the public are all substantially reduced.

A central concern of probability sampling is to avoid bias in the selection of the sample. Bias, or *sampling error*, refers to a discrepancy between the distribution of characteristics in the sample population and the population as a whole. In the first case study, a large university, also its major economic activity, is located adjacent to Middlesville's CBD. If the Middlesville Planning Department wanted to poll people's opinion on downtown parking, they could have someone interview every tenth person who passed by a particularly busy street corner or mail a questionnaire to every tenth person on the city's list of property owners (who pay the city's bills). More than likely, the street interviewer would conclude that there was enough parking downtown (though not enough bike routes), having talked to quite a few university students who did not have cars, whereas one could expect opposite sentiments from property owners, the majority of whom live farther away from the CBD, partly to avoid the student population, and normally drive to town. Either view is skewed and probably would not be an accurate reflection of how Middlesville residents (students, home owners, and others) felt overall.

Sampling frame

Selecting a sample is made from some form of *sampling frame*. Those used most frequently are lists, registers, and maps, either singly or combined. The essential requirements are that the frame must cover the entire population, must be complete (exclude no one), avoid duplication (include no one twice), be accurate and up to date, and be accessible and available for use by the investigator. In the Middlesville parking example, the city tax assessor's list would not qualify by itself because it does not include residents who rent property. If any of these factors is lacking, the situation should be remedied before sampling begins. Given the constraints of fixed schedules of budgets, it may be more feasible to alter the survey design to compensate for inadequacies in the sampling frame. In the Middlesville case study, planners may believe that the revitalization plan requires mapping the location and characteristics of families with school-aged children living in the study area. They may already have decided to use an unweighted cross-section survey design. A sample can then be obtained by selecting every *n*th family from the files of the school board—the sampling frame. Realizing that planning educational facilities requires a lead time of several years, the planning department also has decided to supplement the school board files with the public health department's birth registration lists of the past five years to account for families with children not yet in school.

The sampling design in which every *n*th family on a list was chosen is known as *systematic sampling.* (See Figure 2.5.) To guard against any human bias in using this method, the first unit is selected randomly. (This section on sampling designs should not be confused with the earlier discussion on survey designs. The former is limited to methods of selecting the sample to be used in the survey, whereas the latter refers to formulation of the overall survey design to achieve specified objectives.)

An even more elementary form of probability sampling is the *simple random sample.* The most important criterion in selecting a sample by simple random sampling is that each unit has an equal chance of selection. This can be achieved by two basic methods, variants of each other. In the first case, each unit is assigned a number; these are well mixed, and numbers are drawn from the pool one by one until the sample reaches a size determined previously. In the second case, units are again assigned numbers; however, this time the numbers selected correspond to the numbers on a table of random numbers. Tables of random numbers are usually generated by a computer and often can be found as an appendix in statistics textbooks. (See Figure 2.6.)

Both the simple random sample and the systematic sample use a single stage. Most surveys, however, especially area surveys, make use of some form of

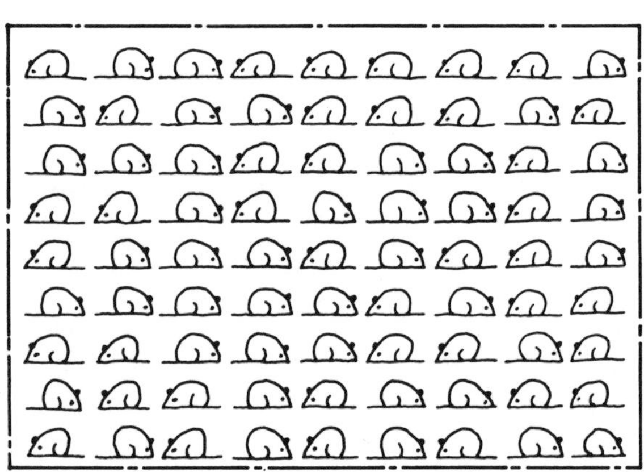

GIVEN POPULATION

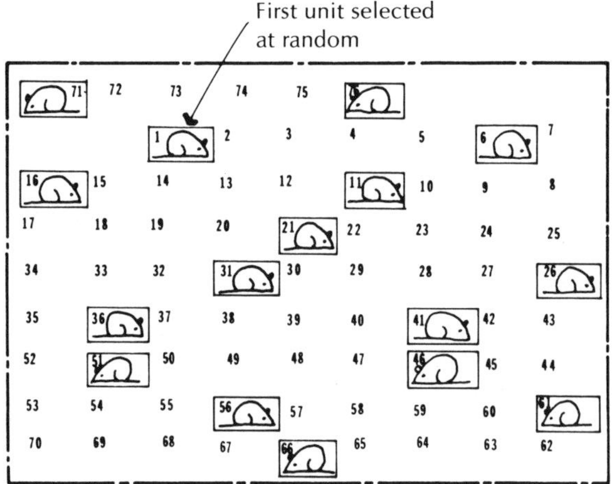

SAMPLE

Every N^{th} (5th) unit selected, first unit at random

Figure 2.5 Systematic sample.

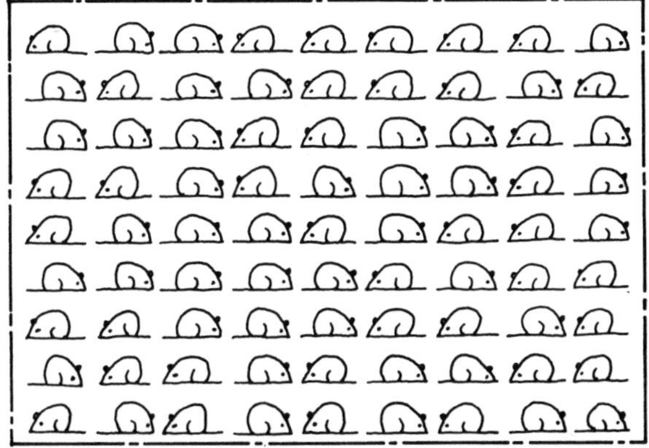

GIVEN POPULATION

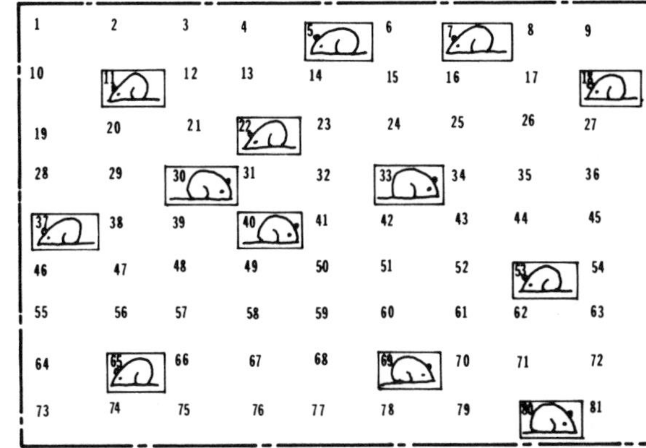

SAMPLE

a) Predetermined number (13) taken at random
b) Predetermined number (13) taken using random numbers

Figure 2.6 Simple random sample.

multistage sampling. Two commonly used forms of multistage sampling are stratified sampling and cluster sampling.

In *stratified sampling* the total population is first divided into subpopulations called *strata*. Each sampling unit is then placed in one (and only one) stratum. The first stage of the sample includes all strata. Therefore if a survey is to be conducted in neighborhoods, each housing unit could be categorized into a particular census block, the stratum. All census blocks would be included in the first stage listing. In the second stage, a sample of individual units would be selected randomly from within each census block. (See Figure 2.7.)

The process is inverted in *cluster sampling*. In this case, the researcher chooses a sample of census blocks in the first stage. Subsequently data would be collected from all units within a census block. Thus, there is a denser sample within selected sample blocks. (See Figure 2.8.)

Sample size

The size of the sample is another matter of choice and is based upon several factors. Primarily sample size depends on the homogeneity of the population being studied. *Homogeneity* is the degree to which units are alike with respect to the characteristics being measured. A population that has a greater degree of similarity can be represented with a smaller-sized sample. Thus it is necessary to have some estimate of the degree of homogeneity (or heterogeneity) when establishing the sample size. When such data are already available, perhaps through a pilot survey, census material, or other secondary source, one need not have as large a sample as when little is known about the population.

Sample size also is related to the sampling method chosen. In general stratified sampling calls for the fewest number of cases; however, this presumes sufficient information to stratify effectively. With less

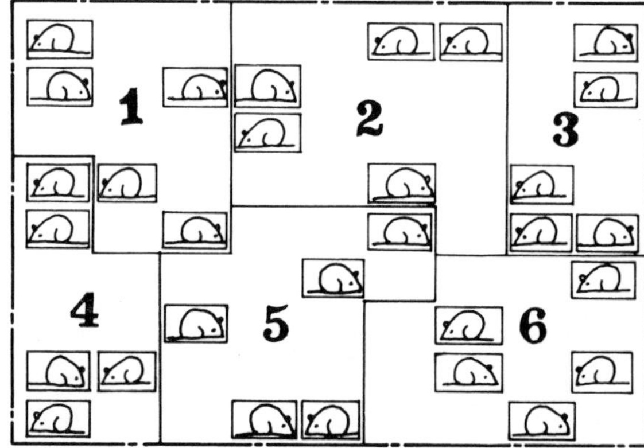

GIVEN POPULATION

SAMPLE

Stratified by census tracts, income, age, or other category
N (5) individuals sampled in each stratum

Figure 2.7 Stratified sampling.

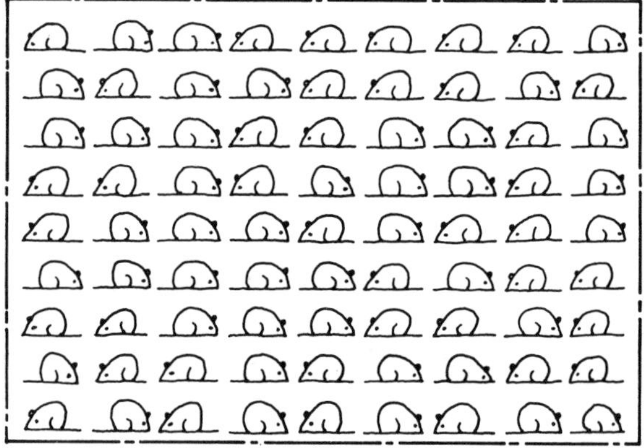

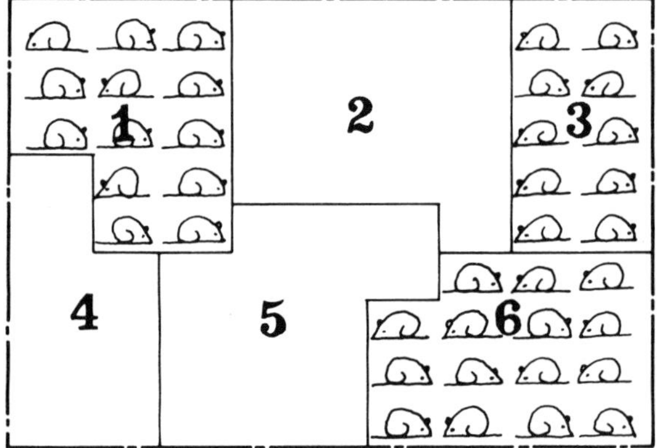

GIVEN POPULATION

SAMPLE

Survey all units in selected categories (census tracts 1, 3, and 6)

Figure 2.8 Cluster sampling.

descriptive data, the simple random survey is recommended even though it requires more cases. Cluster sampling calls for at least as many cases as, or more cases than, the simple random survey.

The way in which collected data will be used in the analysis phase also affects sample size. The more categories by which the data are to be analyzed, the larger the sample needed. Information output will be eroded if a complex analysis yields categories that contain few or no cases. One general rule is no fewer than twenty cases per cell, but even this may be insufficient when there is a lot of variability among units in the sample and when the total number of cases is small. As an exercise during the survey planning phase, the survey administrators together with the report writers and other data users (if these are different people) should attempt to construct the data tables that will be presented to verify that information about the desired cases indeed can be collected.

Finally cost considerations and the time and personnel available for survey work have direct consequences for sample size. The task of juggling these factors will be even more complicated when the survey represents only one component of an organization's overall activities and allocation of resources.

Sample error and confidence levels

Quantity is not a substitute for quality. Sample size does not eliminate error, and a sample does not gain in accuracy when the numbers are increased if the selection procedure is itself biased or faulty. Sampling error built into any sample design affects the reliability of all subsequent estimates and thus impairs the generalizability of survey conclusions to the total population. By employing probability sampling at all stages the researcher has greater control over the quality of results. Statistical theory in effect identifies the level of error attached to a given sample design. The researcher is responsible for deciding how tolerable that error is.

An investigator often uses another statistical tool to establish scientific rigor, the *confidence level*, which specifies the degree to which he or she is confident that the estimate obtained from the sample is correct. At a 95 percent confidence level, the researcher is willing to accept the probability of being wrong five times out of one hundred.

Nonprobability Sampling

Although probability sampling is regarded as a superior method, the planner should not completely rule out the use of surveys by nonprobability methods. When the cost of probability sampling is prohibitive or if representativeness is not necessary, nonprobability sampling may be an efficient course of action and pursued as assiduously as one would a scientifically rigorous methodology. Four commonly used types of nonprobability sampling methods are availability sampling, purposive or judgmental sampling, quota sampling, and volunteer sampling.

One of the least expensive sampling methods to employ is to stop people passing a certain location or to survey people already assembled as a group. The emphasis here is on subject availability and convenience for the interviewer. Survey respondents may be chosen haphazardly, or it may be desirable to stop the first *X* number of people. The latter is sometimes called *chunk sampling*. These techniques would probably be only slightly more sophisticated than field interviews conducted as part of site reconnaissance. While they can yield some interesting insights and colorful quotations to perk up the report, such survey findings should not be used to make general statements relating to the population as a whole.

In *purposive or judgmental sampling* the researcher relies on his or her own acquaintance with the population and its characteristics. One use of this sampling technique is to select the widest variety of respondents possible to identify inadequacies in the survey instrument such as questions subject to different interpretations by different subgroups in the sample. This is more accurately a pretest rather than a survey. Nevertheless it illustrates the more limited purposes for which such data would be used.

Quota sampling begins by constructing a matrix in which the rows and columns are categories of descriptive characteristics relevant to the target population. Each cell is assigned a sample amount in proportion to the number of persons in the total population possessing those characteristics. The researcher then surveys persons having all the characteristics of a given cell until the quota is filled. Quotas may be defined loosely. For example, interviewers simply may be instructed to find and interview fifteen white men and ten nonwhite men and fifteen white women and ten nonwhite women. As a further check the information can be assigned a weight coinciding with its portion of the total population when analyzed. On the other hand, the matrix may be constructed more carefully by using profile information to determine cell distributions. The effectiveness of quota sampling lies in the accuracy of the initial matrix and minimizing any biases in the selection of persons sampled within a cell. Stratifying variables should be chosen carefully because certain characteristics such as sex, age, and race are more readily apparent than others such as income, occupation, and level of educational attainment.

Volunteer sampling involves a self-selected group of survey participants as a matter of course. It is commonly used in surveys of radio audiences and magazine readership polls. People who respond tend to have high levels of interest in the survey subject. The planner may find it beneficial to use this sampling method to get a sense of strong public reactions, whether positive or negative.

DATA GATHERING METHODS

Prior to constructing the questionnaire, another decision must be made regarding the research strategy: the mode by which the survey is conducted. The four traditional methods of gathering data in sample surveys are the face-to-face personal interview, the telephone interview, the mail-in questionnaire, and the self-administered questionnaire completed in a formal group setting.

Face-to-Face Personal Interview

The interviewer-administered questionnaire method is supported by several strong advantages. First, in such a situation, the interviewer can better arouse the respondent's interest in the study and thereby increase the chance that an individual will participate. Second, by creating an atmosphere conducive to discussion, the interviewer can often motivate the respondent to provide complete and accurate answers. Third, the interviewer can avoid circumstances under which the respondent would have decided to skip a question. A skilled interviewer can clarify vague responses, repeat questions, and make questions or words more intelligible to the individual (without influencing the response given). An interviewer on the spot can probe for more detail when a response seems irrelevant or incomplete. Furthermore the interviewer can note characteristics, such as the respondent's appearance, living area, and reaction toward the study. The opportunity for interviewers to observe a respondent's nonverbal behavior should not be overlooked. Fourth, an interviewer has greater control over sequencing questions, often a critical aspect of the survey instrument. Fifth, employing interviewers can often increase accessibility to members of the population who are relatively isolated or physically less mobile. Finally, responses can be enhanced through the use of such visual aids as maps, diagrams, and photographs. The often-heard phrase has many applications in surveys; a picture sometimes can communicate an idea more quickly and precisely than word descriptions, which require that the respondent construct a mental image.

Against these advantages must be weighed the high costs of time and travel expenses and the possibility that an interviewer will distort a response or inadvertently inject his or her own opinion into the phrasing or emphasis of a question or the recording of an answer.

Telephone Interview

This method has many of the advantages of the face-to-face interview with the additional advantage of greater economy. Results from telephone surveys can usually be obtained in a shorter time than personal interviews and mail surveys, and the marginal cost of covering a larger geographical area is not much greater. Telephone interviews can avoid some of the physical risks to interviewers of working in high-crime areas, which is exacerbated by a need to work after dusk when a majority of the sample population is likely to be at home. And respondents who are fearful of opening their door for a stranger may be willing to talk at length to the same person on the telephone.

The advent of random digit dialing has been touted as a means of getting around a major sampling problem: incomplete sampling frames. Nevertheless telephone interviews have their own disadvantage. One of the most significant limitations of the telephone interview is the difficulty of establishing the same degree of social rapport and interaction between interviewer and respondent that is possible in face-to-face situations. The respondent can terminate the interview by hanging up. As a result questionnaires used are usually shorter and less demanding than those used in other data-gathering methods. In some areas, residents may be biased against telephone surveys if they have been widely used as lead-ins to a sales pitch or other types of solicitation by phone.

Self-Administered Questionnaire

The mere presence of a researcher to administer the questionnaire can be sufficient to motivate persons to participate in the study, while the relative anonymity of the self-administered questionnaire makes it a favored mechanism for treating highly personal or potentially embarrassing issues. A person may hesitate to answer such questions verbally although be willing to write out responses, especially if there are no visible markings to connect the respondent with his or her questionnaire form. Self-administered questionnaires also permit the use of graphically designed items, as well as the use of rating scales, checklists, and other forms of measurement too unwieldy or time-consuming for the interviewer to present verbally.

A major drawback of this approach is lack of control over the way in which an individual answers questionnaires or of the quality of those responses. Respondents can and frequently do look ahead, skip around, and compare their answers with those of others.

Mail-in Questionnaire

By far the greatest advantage to this strategy is its low cost. Mail-in questionnaires also benefit from the fact that they allow participants to complete the forms at their leisure. Under some circumstances it may be desirable to give respondents an opportunity to check their responses against other records before writing them down or to confer with others. Mail surveys, however, are notorious for their low response rates. Two or more mailed reminders to respondents could spread the project over several weeks, and even then survey mailbacks may not coincide with the researcher's schedule.

Generally a response of at least 50 percent can be considered adequate for analysis and reporting, a response rate of at least 60 percent is good, and a response rate of 70 percent or more is very good. These rough guides have no statistical basis, however,

and a lack of response bias is more important than a high response rate (Babbie, 1973).

Some general observations can be made about the perennial problem of increasing the mail-in response rate. The greater the amount of work required of the respondent, whether in the form of a long questionnaire or the need to look for an envelope and attach postage, the lower the response rate. Some type of personal touch may be helpful, such as the use of first-class mail, unusual stamps, colored stationery, or a personally typed introductory letter (not too difficult with the help of word-processing equipment). Other more gimmicky devices have been used successfully, such as the promise of a state lottery ticket in exchange for a completed questionnaire. Finally, the greater the intrinsic interest of the subject matter to respondents or the greater the links between the researchers and respondents, the larger the response rate.

Any of these four survey strategies can be used in combination. One of the most successful uses of telephone interviews has been in follow-ups of face-to-face interviews. Similarly a telephone call can be made to remind potential respondents to return their mail-in questionnaire, or researchers can personally deliver the mail-in questionnaire (and explain its purpose) and/or pick up completed questionnaires. Even if people cannot be persuaded to participate in the study, it may be possible at least to obtain some profile characteristics so they may be compared with the characteristics of people who did respond to check for nonresponse bias. (Alternatively nonrespondent bias in mail surveys is sometimes assessed on the premise that nonrespondents are more similar in characteristics to people who responded after follow-up efforts rather than to early respondents.)

DESIGNING THE QUESTIONNAIRE

The common element in the data-gathering methods discussed above is their use of a questionnaire. Researchers have several options in designing the exact format of the questionnaire, primarily in the construction of questionnaire items and sequencing of questions.

Constructing Questionnaire Items

Two basic categories of questions are the open- versus closed-ended inquiries. In the *open-ended question*, the respondent comes up with his or her own answer and therefore has considerable freedom in relating to the question. With *closed-ended questions*, the respondent is asked to select a single answer from a list provided by the researcher (that which "best fits") or multiple answers (those that "apply").

Closed-ended questions are useful when the respondent is asked to make distinctions of degree. The *rating scale* is one technique and provides an ordinal measurement of degree. Commonly used is the *semantic differential*, which presents a set of opposite word pairs and gives the respondent seven degrees of response:

"How would you usually describe the street activity outside your house or apartment?"

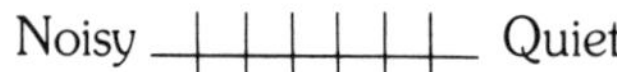

Another popular scale, the *Likert scale*, calls for the respondent to indicate the extent to which he or she agrees with a statement:

"There are adequate parking facilities in the major business district on Main Street between Second and Fifth streets."

Strongly Agree Agree Disagree
Strongly Disagree Undecided

Choices do not have to be indicated by words. In one case, an urban vest-pocket park was designed successfully with help from citizens who completed picture ratings of preferable park layouts. The pictures, actually photographs of simple models, were a highly suitable communication medium. Not only were the pictures engaging, thereby inviting people to participate in the survey, but they also provided sufficient imagery so that persons not trained in design could work on the problem (Kaplan, 1978).

The responses prompted by open-ended questions are usually more difficult to code and analyze because they do not fall neatly into pre-determined categories. However, lack of structure, an opportunity for self-expression, and spontaneity can bring important new insights to a study if researchers are willing to put up with human idiosyncrasies.

The choice of questions used will ultimately depend on the aims of the study, the type of respondents (and their competence to answer particular questions reliably), and the purpose of the specific questions. Some general guidelines apply to all questionnaire items, however. First, any item on the questionnaire should be straightforward and unambiguous. Second, the investigator should not expect respondents to give a single answer to what is actually a combination of questions. Double-barreled questions can be avoided by keeping items short. In any case, it is safe to assume that many respondents will read items quickly or pay less than full attention to the interviewer and want to provide quick answers. Under these conditions, clear, short items can lessen the possibility of misinterpretation. Third, questions should be relevant to most respondents. When attitudes are queried on topics that few respondents have thought about, the results are not likely to be useful unless information about the level of interest is purposefully requested. Similarly, researchers must be careful to avoid *expert error*, the error of attributing to the respondent a degree of expertise which he or she does not possess in the field in question.

Care should be exercised in the choice of words. Basically this problem is similar to the idea of identifying a common vocabulary and frame of reference. Each respondent will be able to understand and interpret the questionnaire items only according to his or her own experiences. Therefore the investigator formulating the items should constantly ask, "What do the questions mean to me?" "What do they mean to the respondent?" In the latter case, profile data such as age, education, sex, and ethnicity can aid the task of estimating the respondents' frame of reference.

Ordering Questionnaire Items

With careful planning, the survey researcher can often work out a questionnaire design that combines a variety of response forms and uses each advantageously. This then becomes largely a matter of patterning questionnaire items. Sequencing questions is equally important because it will affect the amount and quality of information gathered. Of various patterns that can be used, the most basic are the funnel sequence and the inverted funnel. The *funnel sequence* refers to a procedure in which the interviewer first asks the most general, open, unrestricted questions. In contrast, the *inverted funnel* starts with specific questions and then moves to more general issues.

When sequencing questions, the researcher should be sensitive to the logic respondents follow, thereby helping them to provide accurate information. It is especially important that the relationship of a question and the overall purposes of the study make sense to the respondent. Use of transitional questions or explanations can be inserted to show how a new topic relates to what has been discussed previously or to the purposes of the study.

EVALUATING THE SURVEY INSTRUMENT

Many of the specific details regarding design of the survey strategy and instrument will be satisfactorily worked out only after a great deal of discussion, revision, and reorganization. Issues ranging from where to locate sensitive items within the interview schedule to what physical layout of the questionnaire is appealing to the respondent yet functional for the researcher are best evaluated by putting the study into operation and seeing how it works. A preliminary evaluation can be most efficiently accomplished by making use of such tools as the pilot study, pretest, and trial run.

In the *pilot study* the researcher wants to anticipate alternative situations he or she will face under full-scale operating conditions so that contingency procedures for dealing with potential problems can be mapped out in advance. The pilot study can be either exploratory (what happens if we do this?) or estimative (for example, checking various characteristics of the population and/or environment to confirm sample group selection and availability).

The *pretest* is essentially evaluative. It is used to help decide which of the alternative procedures should be used. This requires that the researcher set up criteria of effectiveness beforehand so choices can be made about which alternatives are good enough to be employed in the large-scale study.

The *trial run* is used for a final check of whether all possible alternatives have been considered and the most efficient procedures have been chosen. Thus this study is designed to evaluate the operational plan as a whole before the final run. The trial run is also important as a training tool for all personnel who will take part in the actual survey. In addition to making sure that all materials are available when necessary, those who will use them can familiarize themselves with the materials under realistic conditions. As a result the trial run gives the research and planning staff preliminary information about the adequacy of personnel and operations. They can learn where more or different personnel may be needed, how communications can be improved, and how long it will take from initial contact with respondents until data are ready for analysis. On the basis of concrete information from the trial run, many procedural problems can be ironed out and last-minute adjustments made to the financial and/or time budgets.

ALTERNATIVES TO THE SURVEY

Observations

One of the investigator's greatest fears throughout the process is that his or her work will meet with resistance from the population being surveyed. An available option in such an eventuality is observation. For example, retail outlets in the CBD may refuse to be surveyed because they suspect competitors will be alerted to their business conditions. However, the investigator can still observe the activity of delivery vehicles or the number of employees and customers arriving and departing at various times. Like the survey methods, an observation schedule must be carefully designed to minimize inaccuracy and bias. The observation and recording of data can be aided with a list of required information devised before the field survey begins. In this way the observer knows where to concentrate his or her attention when confronted with a barrage of sensual stimuli.

Every item must be fully recorded; reliance should never be placed on memory. Moreover, recorded information should distinguish between observations of actual occurrences and interpretations. To resolve the problems of recording events simultaneously with their spontaneous occurrence, sound-recording equipment should be used when feasible (the decreasing cost of operating video-recording equipment is making this technology increasingly attractive) or two or more people should observe the same events.

Key-Informant Technique

The key-informant technique has been used most extensively by anthropologists studying the structure and behavior of cultures. It refers to a method of information gathering by tapping the knowledge of a few people, usually through unstructured personal interviews. These people occupy positions or roles that allow them to communicate a broad, synthesized picture about a certain subject and/or a specialized picture of that topic. Use of key informants is appropriate when the objective is to obtain comprehensive or in-depth information not expected from a sampling of the population. Also appealing are the low costs.

Before engaging in this technique, the investigator must consider the qualifications of proposed key informants to decide whether each possesses the relevant information. Consideration should also be given to the sources of any biases the informant may have, possibly as a result of that unique role. This problem may be controlled somewhat by using multiple informants. In any case, the key-informant technique is not meant to take the place of survey data with its emphasis on unbiased estimates for projection to a more general population.

Key informants are especially valuable where there is a communication gap between the target study population and the researcher. Individuals are selected for their ability to bridge the gap by speaking the language of both sides. Often the interviews will be held as a preliminary phase of a research project to see that the right problem and critical variables have been identified, somewhat as in a pilot study. (A method for identifying opinion leaders, one type of key informant, is discussed in Chapter 6.)

Focus-Group Interviews

Much of the work on focus-group interviews has been in the area of marketing research. Basically this technique involves group interviewing; however, rather than asking questions of each member in turn, the moderator stimulates discussion, which becomes a combined group effort. With several members participating, each person alone is not responsible for coming up with an answer, which can reduce the level of anxiety present in personal interviews and make for more thoughtful responses.

A major asset of this technique is that one person's comments can spark new ideas in another person, thus setting off a chain reaction. The group forum also allows people to explore more thoroughly similarities and differences in their experiences.

On the whole, this method is inexpensive, although cost can vary depending on the participants. Focus groups are fairly easy to organize. The difficulty lies in finding a skilled moderator and in analyzing the results of the session. How do you judge one person's comments against another's? What effect did the group's environment have on the statements made? This kind of qualitative research can provide useful information about human experiences not found in numbers, but one must work with the inherent limitations of analyses based on subjective interpretation. With or without follow-up with quantitative research, the investigator has a professional responsibility to report accurately whatever method was used to derive stated conclusions.

CONCLUSION

Despite the plethora of data stored in computer banks or in hard-copy files, planners will still rely on surveys as one of the basic methods for collecting information pertinent to their activities. This chapter has outlined the major components of survey research, some available options, and factors to consider in putting together a project. It has been emphasized that the specific nature of a survey project will depend on the objectives being sought and the resources available for such an endeavor. Successful completion of a survey calls for adequate strategic planning and general attention to tactics and the ability to maneuver in midstream. The survey is a tool that will continue to serve diligent and creative planners well.

APPLICATIONS

1. We reenter the scene in Middlesville with the planning department in the midst of revising a comprehensive revitalization plan for the CBD and surrounding neighborhoods. The city planners believe that adequate information is not available on public needs and preferences. Junior Planner, realizing that the planning process would benefit greatly from direct citizen participation, proposes that a survey be conducted. He even suggests a title for the study: "Middlesville Tomorrow: Toward the Year 2000."
   a. Draw up a list of objectives for such a survey. Begin by asking questions you would like the

survey findings to answer. For example, you may want to investigate the relationship between a vital CBD and a high-quality urban life-style. Compare your list with that of one or two other persons and try to resolve any differences in survey objectives.

b. Now draw up a list of objectives that a neighborhood group, Citizens Against Towers (CAT), would adopt if it were to conduct a separate survey.

2. After consulting a sampling expert and weighing administrative costs, the city's planning committee for the survey decides on mail-in questionnaires as most appropriate.
   a. Prepare a cover letter to send with each questionnaire. The primary objective of the letter is to encourage participation and thus minimize the non-response rate.
   b. CAT, an organization comprised of volunteers, feels that face-to-face interviewing is a more effective strategy. Write out an introduction to the interview that can be presented verbally.
3. At one point, the possibility of distributing the city's questionnaire as an insert in the only local newspaper was considered. What are the pros and cons of this approach? How would your assessment differ if newspaper officials informed you that the Sunday morning edition was delivered to 85 percent of the households in Middlesville? How would it differ if an address list were also supplied?
4. Middlesville planners are interested in learning how attractive the public finds CBD retail facilities and whether there is support for a downtown shopping mall favored by the Merchants' Association. Despite the importance of this issue, questionnaire items related to it can fill only a limited amount of space. Construct five items that would yield usable information. These can be closed- or open-ended questions, scales, or graphic options.
5. CAT has developed a simple sampling procedure of sending an interviewer to every fifth house on the block. Organization members believe that probability sampling should be conducted to the degree feasible. In addition to a small budget, they face stringent deadlines to develop counterproposals to the city's CBD plan. Under the circumstances, interviewers have been instructed to do their best when unexpected situations arise.

   If you were an interviewer and found a group home where the map indicated a single-family dwelling, what would you do? If you were the survey project director, how would you use this contingency to refine the sampling design?

   Contact a person with previous survey interviewing experience and find out how this and other unusual situations were resolved.
6. One alternative to surveys discussed in the chapter is the focus-group interview. To practice this technique, set up a group of four or five persons with one person as interviewer. After all group members have read the same news account of a prominent local event, the interviewer should find out what the other group members think about the subject. Spend about twenty minutes for this discussion. Then repeat the process, but this time have group members read and discuss an editorial.

REFERENCES

Babbie, Earl, 1973, *Survey Research Methods,* Wadsworth, Belmont, Calif., p. 165.

Kaplan, Rachel, 1978, "Participation in Environmental Design: Some Considerations and a Case Study," in Stephen Kaplan and Rachel Kaplan, eds., *Humanscape: Environments for People,* Duxbury Press, North Scitaute, Mass., pp. 427-438.

Reingen, Peter H., and Jerome B. Kernan, 1978, "Compliance with an Interview Request: A Foot-in-the-Door, Self-Perception Interpretation," in Robert Ferber, ed., *Readings in Survey Research,* American Marketing Association, Chicago, pp. 541-549.

Sudman, Seymour, 1967, *Reducing the Cost of Surveys,* Aldine, Chicago, pp. 68-88.

BIBLIOGRAPHY

General Texts on Survey Methods

Babbie, Earl, *Survey Research Methods,* Wadsworth, Belmont, California, 1973.

A good, comprehensive treatment of survey methods with practical tips on how to avoid pitfalls from the author's experience. See especially the sections on the survey as a research tool.

Backstrom, Charles H., and Gerald Hursh-Cesar, *Survey Research,* 2nd ed., John Wiley, New York, 1981.

The expanded second edition contains some of the recent developments in survey research and is presented in an easily readable style with many helpful checklists.

Warwick, Donald, and Charles Lininger, *The Sample Survey: Theory and Practice,* McGraw-Hill, New York, 1975.

An excellent introductory textbook on the development and administration of surveys. Examples are used liberally throughout.

Works of Particular Relevance

Bellenger, Danny N., Kenneth L. Bernhardt, and Jac L. Goldstucker, *Qualitative Research in Marketing,* Monograph Series 3, American Marketing Association, Chicago, 1976.

Although the applications tend toward marketing, there is a good discussion of the benefits and limitations of using non-quantitative data-gathering techniques.

Dillman, Don A., *Mail and Telephone Surveys: The Total Design Method,* John Wiley, New York, 1978.

Step-by-step detailed methods for conducting mail and telephone surveys with explicit attention to techniques that can maximize respondent participation.

Hyman, H., *Secondary Analysis of Sample Surveys: Principles, Procedures, and Potentialities,* John Wiley, New York, 1972.

A thorough examination of the uses of survey data that have already been collected.

Kish, Leslie, *Survey Sampling,* John Wiley, New York, 1965.

A standard text providing comprehensive coverage of sampling procedures.

Social Science Research Council, *Basic Background Items for U.S. Household Surveys,* Social Science Research Council, New York, 1975.

A suggested set of basic background questions and coding procedures for general population household interview surveys with explanatory and supplementary notes and references to relevant research literature.

Sudman, Seymour, and Norman M. Bradburn, *Response Effects in Surveys: A Review and Synthesis,* Aldine, Chicago, 1974.

Good treatment of the mechanics of interviews. Attention is given to the interactions between interviewer and respondent and the biases that could arise from such interactions.

University of Michigan Survey Research Center, *Interviewer's Manual,* rev. ed., Survey Research Center, Ann Arbor, Mich., 1976.

Recommended as a practical guide for interview situations.

Journals

The major source of articles concerning the development and administration of surveys is the *Public Opinion Quarterly.* Other journals containing similar articles, but in fewer numbers, are *American Sociological Review, Journal of Applied Psychology,* and *Journal of the American Statistical Association.*

3

Information from Secondary Sources

Allan Feldt

A substantial amount of information relevant to many planning problems has already been collected by other organizations. Such information is often fairly easily available in published form or on machine-readable tapes, discs, or cards and can be found in public or university libraries or in the files and records of private businesses and various public agencies. Finding and using such information enables a planner to develop quickly relatively inexpensive background information and analyses on particular planning issues and sections of the city.

Two major problems exist in using such sources. First, such information has usually been collected for some purpose other than the needs of the local planning office. Knowledge of what kinds of data are available together with imaginative analysis can usually produce some useful information on almost any issue, however. Another problem with secondary sources is created by the enormous diversity of types and sources of information and the associated problems of choosing and gaining convenient access to them. By providing an overview of the major types of such information, this chapter attempts to enable planners to use secondary data sources intelligently and selectively.

A chapter such as this that deals only with sources of information cannot be understood or utilized through a direct reading. I recommend that it only be skimmed at this time so that its general content and organization are understood. Then when specific information is sought at some later time, reference to the specific section of this chapter appropriate to your needs can be made without your having to review the entire contents. Keeping copies of three or four basic publications available in the office can be very helpful in providing an immediate source of some aggregate statistics and references to additional sources of information. Particularly useful in this regard are recent copies of the *County and City Data Book*, the *Bureau of Census Catalog*, the *Statistical Abstract of the United States*, and the *Municipal Yearbook*.

Having read this chapter several months previously, the Junior Planner trying to solve trash disposal problems on Main Street would have a vague recollection of several possible sources of information on levels of operation of other municipal governments. Thinking of three or four comparable cities to Middlesville, he could have used an office or library copy of the *County and City Data Book* to compare the general revenues, educational expenditures, and total outstanding debt of these communities to those of Middlesville. A recent copy of the *Municipal Yearbook* would provide additional information on expenditures and levels of services offered by similar cities, as well as the names of current city officials if he should wish to contact them directly for further information. A special article or review of solid-waste-disposal techniques in small cities could probably be found in a copy of the *Municipal Yearbook* published within the last five years. The *Census of Governments* would provide additional information on local government organization, employment, and finances, which might also provide useful background information for better understanding the situation in Middlesville.

The information and references provided in this chapter refer only to U.S. publications and programs. The enormous diversity of materials and procedures found in other countries precludes their coverage here. The procedures and publications of the U.S. Census Bureau have been adopted in many other countries, however, and familiarity with the materials described for the United States provides a useful background for finding and using similar materials elsewhere. While information collection and dissemination procedures in the United States are more elaborate and detailed than those found in many other countries, the United States provides information of lower quality in at least two areas. The Census of Population has been conducted decennially; thus there are prolonged periods during which no current information on population size or characteristics is available. Several other countries have been conducting population censuses every five years for some time, and as a result both the quality and currency of their demographic information are superior to that of the United States.

Second, many other countries maintain continuous registration of their population, requiring individual citizens to inform the government of significant changes in their demographic status, including any change of address. When such permanent population registers exist, demographers and planners find their understanding and analysis of population changes and movements much improved. No such registration system exists in the United States, and it is unlikely that one will be created. While population registration systems are useful to planners and demographers they are also useful to police and intelligence agencies. It seems likely that potential threats to personal liberties will continue to be more important than the advantages of establishing a permanent population register.

THE CENSUSES OF POPULATION AND HOUSING

The U.S. Bureau of the Census conducts ten regular censuses plus a wide variety of other activities of value to urban planners. The most widely known of these censuses are the Census of Population and the Census of Housing, both of which have been conducted once each ten years.

The Decennial Census of Population collects information on sex, race, age, marital status, Hispanic origin, and household relationship of each person in the country. Additional information is collected on a sample of individual households to provide data on education, ancestry and place of birth, migration, marital history and number of children, employment status and place of work, means of transportation to work, and income.

The Census of Housing is conducted simultaneously with the Census of Population and collects

information from every household on the number of dwelling units present, the number of rooms present in each dwelling unit, the condition of plumbing facilities, and a variety of data on ownership, rental status, and value of the dwelling unit. As in the Census of Population, a sample of households is asked additional questions, providing information about their buildings such as number of stories, sources of water, sewage disposal, methods of heating and cooking, number of bedrooms and bathrooms, presence of telephones and air conditioning, number of automotive vehicles present, and total shelter costs, including mortgages, taxes, and insurance.

Information from these two censuses is reported in a wide range of publications and is also made available on computer tapes and microfiche. Additionally special tabulations may be ordered at cost from the Census Bureau for some information that has been collected but not published. Although somewhat expensive, special tabulations provide data to local governments at a fraction of the costs that would be incurred by collecting the data locally. The major form of publication is the *Census of Population* in four separate chapters issued sequentially for the entire country and finally bound in a single volume for each state. The first chapter provides information only on the number of inhabitants; later chapters provide increasingly detailed information on population characteristics, social and economic characteristics, and further detailed characteristics on larger population centers and aggregates.

Data in the first two chapters on number of inhabitants and general population characteristics are reported for virtually all identifiable civil divisions down to and including unincorporated places of 1,000 or more population. Data in the third and fourth chapters are reported only for cities of 10,000 or more and for counties and states. Detailed data in Chapter 4 are usually reported only for larger cities and Standard Metropolitan Statistical Areas (SMSAs). Some data, including data from the Census of Housing, are tabulated for individual city blocks and for census tracts and are published in census tract and in block statistics volumes for the SMSAs and for a few smaller cities that made prior arrangements to pay for the special publication of block statistics. The only data on individual households made available or published are those of an anonymous national sample of households, for research purposes.

Other U.S. Censuses

In addition to the Census of Population and the Census of Housing, the Census Bureau conducts a number of other censuses on various forms of economic activity every five years. These include the censuses of manufactures, wholesale trade, retail trade, service industries, agriculture, construction industries, mineral industries, transportation, foreign trade, and government.

Data reported in these economic censuses concentrate on the size, volume, and operating characteristics of the businesses being covered. Total receipts for sales or services, number of employees, square feet of space occupied, payroll, year-end inventories, operating and capital expenditures, types of organization, and other detailed information unique to each type of business are collected and reported.

Reporting information for these censuses is mandatory, but the Census Bureau protects the confidentiality of information from individual firms by refusing to publish or make available information for small areas when the small number of firms of a given type makes it possible to deduce information from a single firm. The general rule is to withhold data for any area where fewer than four businesses of the same type are included in the numbers reported.

Most of these data are published in volumes for separate states or geographic areas, as well as in volumes for specific industries or types of establishments. Other series on specific subject matter are published appropriate to each of the types of censuses. Most data are also available on computer tapes or cards.

Monthly or annual surveys are conducted for most of these types of activities, and results of these surveys are reported regularly in a series of monthly or annual publications, including *Current Industrial Reports, Annual Survey of Manufactures, Quarterly Summary of State and Local Tax Revenues, County Business Patterns, Monthly Survey of New Construction, Annual Housing Survey, Annual Retail Trade Survey,* and *Current Population Reports.*

Other Census Publications

The Census Bureau issues a number of other useful publications. Following each of the major censuses, a number of special reports and monographs are issued that provide detailed analysis and interpretation of data collected in the census. Such analyses are intended to cover certain specific and timely issues of wide general concern to the public and are sometimes written by academic or private researchers under contract with the Census Bureau. Topics range from an analysis of growth and change in the New England shoe industry to an analysis of fertility patterns among persons of Spanish surname. The *Bureau of Census Catalog* provides a list and brief description of monographs currently available.

The *County and City Data Book*, published every four years, provides some data from the most recent economic censuses and from the most recent censuses of population and housing for cities of 25,000 or more population, for all counties and SMSAs, and for the states, regions, and divisions of the United States. Data presented in this volume are also available on computer tape. This book is probably the most generally useful census reference for providing basic data on most cities and all larger areas of the country. It also provides a ready source of most census definitions and additional sources of information.

The *Statistical Abstract of the United States*, published annually, is the second most generally useful compendium of information for the country as a whole. This volume provides a substantial amount of information on the social, economic, and political organization of the United States, although it includes relatively little information on individual cities, counties, or states. Probably most useful to a local planner is its listing of hundreds of other sources of information, with some indication of their reliability and utility. For this purpose alone, the *Statistical Abstract* is a worthwhile bibliographic source. Since many of the statistics reported change infrequently, purchasing copies more often than every three or four years may not be necessary.

The Census Bureau publishes a number of catalogs and guides to census materials that are helpful in locating particular items and sources of data. In addition to the *County and City Data Book* and the *Statistical Abstract of the United States,* the *Bureau of the Census Catalog* issued monthly, quarterly, and annually and the *Directory of Federal Statistics for local Areas* and its *Urban Update* are especially helpful. Other guides and catalogs to specific censuses, types of data, and uses of census statistics are also provided in such profusion that one needs a guide to census guides. A reference librarian can be of great assistance, although the

profusion of publications is so great that only highly specialized librarians are fully conversant with what is available. The Census Bureau has regional user service centers and provides telephone hot lines that are helpful in some but not all regions. Beyond the statistics provided through the Census Bureau, almost every other major federal agency and many comparable state and private agencies provide thousands of additional types of data. The most useful way to cover these additional sources of secondary data is by major types of information provided rather than by the source of the data.

POPULATION AND VITAL STATISTICS

In addition to the national decennial census and the monthly surveys, a few states provide annual or five-year counts of population based either on an actual census or on estimates of local populations derived from partial censuses and counts such as a statewide school census. In most cases, however, local population data between national census years are based upon population estimates provided by the U.S. Bureau of the Census or by estimates provided by either the state planning department or the state department of public health. Such estimates are subject to considerable error, and the level of error increases substantially as time since the last decennial census increases. The introduction of a five-year census will improve this situation considerably. Current population estimates as well as projections are often computed by local planning agencies themselves. Major sources for information on current population estimates and projections from the Bureau of the Census are the *Current Population Reports*, series P-20, Population Characteristics; P-23, Technical Studies; P-25, Population Estimates; and P-28, Special Censuses.

Nationally the *Vital Statistics of the United States*, published annually by the U.S. Public Health Service, is the most comprehensive summary of data on births, deaths, marriages, and divorces in the country. These data are also reported on a monthly and weekly basis in the *Monthly Vital Statistics Report* and the *Weekly Morbidity and Mortality Reports* from the U.S. Public Health Service.

Local planners should be aware, however, that these data are only a compilation of data provided by local county health departments to their state health departments and through them to the federal agencies. Thus the best source of local information should be the local county health office. Unfortunately many local health offices do not keep open and well-ordered files of this information once it has been forwarded to the state agencies. A few visits to the local health department should determine whether such an agency can be relied upon for these statistics directly. Offers of assistance or an expression of interest might result in improvements in local filing practices or at least in providing copies of future reports to the local planning office. It has been our experience that frequently the local clerks or secretaries responsible for compiling and forwarding these statistics are so pleased to have someone express an interest in their work that they are happy to cooperate with efforts to provide better and more timely data in the future.

Other sources of information on population and vital statistics include the United Nations, the World Health Office, the annual report of the Immigration and Naturalization Service, and numerous private agencies such as the Population Reference Bureau, the Social Science Research Council, the Metropolitan Life Insurance Company, other major life insurance companies, individual state and county health departments, and a number of private publishing companies, such as R. L. Polk, Rand-McNally, and the F. W. Dodge Division of McGraw-Hill.

HEALTH AND WELFARE

Information on the well-being of the U.S. population is provided by a broad range of sources, with the U.S. Public Health Service and the Department of Health and Human Services providing the majority of the information. The Public Health Service publishes: *Accidental Death and Injury Statistics* annually, *Air Quality Data* annually, *Local Health Organization and Staffing within Standard Metropolitan Areas* periodically, *Municipal Water Facilities Inventory* every five years, *Patients in Mental Institutions* at irregular intervals, and periodic reports from *The U.S. National Health Survey* describing the results of monthly interviews and health examinations on a nationwide sample of U.S. families. Information from the surveys is reported only for larger population groupings to protect the privacy of informants.

The Department of Health and Human Services, formerly named the Department of Health, Education and Welfare, publishes *Annual Report on Health, Education, and Welfare Trends,* as well as a monthly report, *Indicators.*

Other periodic or frequent reports from other governmental and private agencies that may be helpful include:

Accident Facts, published annually by the National Safety Council.

Consumer Price Indexes published monthly and annually by the Bureau of Labor Statistics.

Social Security Bulletin (monthly) of the Social Security Administration.

Hospitals Guide and *American Medical Directory* published annually and biennially, respectively, by the American Medical Association.

Facts about Nursing: Statistical Summary published annually by the American Nurses Association.

Welfare in Review published monthly by the U.S. Welfare Administration.

Life Insurance Factbook (annually) by the Institute of Life Insurance.

Monthly Labor Review and monthly reports on employment and earnings from the Bureau of Labor Statistics.

Annual Statistical Series of the Welfare Administration.

Yearbook of American Churches, published annually by the National Council of Churches of Christ in the U.S.A..

Unemployment Insurance Statistics published monthly by the Bureau of Employment Security.

Annual Report and *Monthly Review* of the Railroad Retirement Board.

Annual Report on Corporate Pension Funds of the Securities and Exchange Commission.

Handbook of Old-Age, Survivors, and Disability Insurance Statistics published annually by the Social Security Administration.

Annual Report of the Administrator of Veterans Affairs, Annual Report on Child Welfare Statistics, Public Assistance, and their several Statistical Series on various specific welfare and public assistance programs from the Welfare Administration.

A few additional useful sources might be found in the *Annual Report of the American National Red Cross* and the *Annual Directory of the United Community Funds and Councils of America.* Directories, journals, and annual reports of many other concerned groups such as the National Association of Social Workers and the Public Welfare Association may also be useful sources of information, although not usually for local areas.

RECREATION AND LEISURE ACTIVITY

Information on recreation and leisure-time activities and facilities may be found in publications such as:

State Outdoor Recreation Statistics by the former Bureau of Outdoor Recreation, more recently known as the Heritage, Conservation and Recreation Service.

National Survey of Fishing and Hunting of the Fish and Wildlife Service.

Monthly Public Use of National Parks of the National Park Service.

Statistical Survey of Museums in the United States and Canada published occasionally by the American Association of Museums.

Parks and Recreation published monthly by the National Recreation and Park Association.

The *Recreation and Park Yearbook* (quinquennial) of the National Recreation and Park Association.

Occasional reports by organizations such as Resources for the Future, National Association of State Racing Commissioners, National Golf Foundation, and so forth.

TRANSPORTATION AND COMMUNICATIONS

The Census of Transportation, conducted every five years, actually consists of three related surveys: the *National Travel Survey*, the *Truck Inventory and Use Survey*, and the *Commodity Transportation Survey*. The utility of this information is limited for local planning purposes since most of the data are available only for states and regions. The first consists of a sample of households and describes their travel at various times of the year. It includes information such as distance, purpose, mode of travel, and recreational activities engaged in. These results are reported only for the larger states and for larger geographic units, but unpublished data in greater geographic detail may be specifically ordered from the Census Bureau. The *National Travel Survey* provides the published data for larger areas. Computer tapes of individual household responses are available, with all identifying information removed.

Using a sample of truck registrations, the *Truck Inventory and Use Survey* yields information on number of trucks and their characteristics such as body type, load size, products carried, and mileage for geographic divisions and states. Individual responses are available on computer tape, with identifying information removed.

Traffic flow data on the volume of commodities shipped organized by means of transport, size, origin and destination of shipments, and similar measures are collected from a sample of smaller manufacturers and from an analysis of shipping documents from a sample of plants taken from the Census of Manufactures. Findings are reported as shipments among major metropolitan areas in the *Geographic Area Report Series*, by major industrial categories in the *Industry Report Series*, and by type of commodity in a microfiche series entitled *Commodity Report Series.*

The Federal Highway Administration has twice published a *National Personal Transportation Survey*, which reports information on all trips by members of individual households during a specified twenty four-hour period and longer trips taken during the preceding two weeks. Mode of travel, use of public transportation, purpose of trip, and other information are reported on a national basis. The last available report was a survey conducted in 1977. Other sources of information on transportation include:

American Trucking Trends and Motor Truck Facts, published annually by the American Trucking Association.

American Facts and Figures, published annually by the Automobile Manufacturers Association.

Traffic Volume Trends, published annually by the Bureau of Public Roads.

Waterborne Commerce of the United States, published annually by the Corps of Engineers.

Handbook of Airline Statistics, published annually by the Civil Aeronautics Board and their monthly statistical report on air traffic.

Annual and monthly reports of the Interstate Commerce Commission, plus quarterly and annual reports on specific types of transportation.

Additional private reports on a monthly and annual basis are issued by companies such as the American Transit Association, the Association of American Railroads, the Air Transport Association of America, the Lake Carriers Association, the National Association of Motor Bus Owners, and several similar industry-specific interest groups.

Sources of information on communications include the annual and quarterly reports of the Federal Communications Commission and the annual reports of the American Telephone and Telegraph Company and the United States Independent Telephone Association.

HOUSING AND CONSTRUCTION

In addition to the Census of Housing, the *Annual Housing Survey* offers a major source of information on housing characteristics published for many of the larger metropolitan areas. These data are also available on computer tape with individual identification removed.

The Bureau of Labor Statistics occasionally publishes bulletins and reports on new housing starts, cost of housing, and trends in building permit activity. The *Annual Report* and the monthly *Housing Statistics* of the Department of Housing and Urban Development are also important sources for national information and for major areas of the country.

Further information on costs and volume of construction are provided by the monthly *Dodge Construction Contract Statistics* published by the F. W. Dodge division of McGraw-Hill Publishing Company and the weekly *Engineering News-Record* from the same publisher.

Occasional studies and special reports are published by organizations such as the National Bureau of Economic Research, the Social Science Research Council, and the National Association of Home Builders.

EDUCATION

The Department of Education publishes an extensive series of reports on educational activities within the United States. These include monthly, annual, and biennial reports on state school systems, public secondary schools, colleges and universities, number and type of degrees awarded, library statistics for colleges and universities, statistics on public libraries, and so on. The type and frequency of the reports is sufficiently variable that an examination of current listings is necessary to determine appropriateness and availability of information.

Similar listings of public and state-supported educational establishments are published by most states, usually by the state department of education or a similar state agency. The National Science Foundation maintains a roster of scientific and technical personnel and publishes occasional reports on their numbers, training, and activities.

Nongovernmental sources of information on education are published by the R. R. Bowker

Company, the Council of State Governments, the National Catholic Welfare Conference, and the National Teachers' Association.

GOVERNMENTAL ACTIVITIES

In addition to information from the Census of Governments and the *Municipal Yearbook*, the Council of State Governments issues *Book of the States* annually, the Conference of Mayors publishes annual proceedings, and the International City Managers' Association publishes several annual directories and special studies on local government. Several universities have attempted to develop regular surveys and reports on governmental activities in local and metropolitan areas, although these tend to be dependent on available financial support. Notable among these has been a series on metropolitan planning activities and studies published by the Graduate School of Public Affairs at the State University of New York in Albany.

Many states issue some form of state yearbook or manual giving information on the principal officers in state and local government. Such sources sometimes provide election statistics from recent local, state, and national elections.

The Advisory Commission on Intergovernmental Relations provides a number of studies on taxation efforts and characteristics of state and local governments. The Bureau of Prisons and Federal Bureau of Investigation issue annual reports on crime, arrests, and imprisonment for state and local areas. Moody's Investor Service issues an annual publication with semiweekly supplements on local and state government activities for potential investors in state and municipal bonds. Occasional studies of governmental activities are published by organizations such as the National Bureau of Economic Research, the University of Michigan's Survey Research Center, and the Republican and Democratic National Committees.

ENVIRONMENTAL AND NATURAL RESOURCES

Information about local and regional climatic factors is issued on weekly, monthly, and annual bases by the Environmental Science Services Administration. Information on forests, fisheries, and wildlife is provided monthly and annually by the U.S. Fish and Wildlife Service and the U.S. Forest Service. The Bureau of Mines provides data on mineral, coal, and petroleum production in a series of monthly, quarterly, and annual reports. Similar information is also provided by private industry sources, such as the American Petroleum Institute, the National Coal Association, the American Bureau of Metal Statistics, and the Commodity Research Bureau.

Various agencies within the present Department of Energy issue annual reports on nuclear energy sources and activities, on electrical energy generation, on natural gas production and consumption, and on a number of other energy resources. Similar reports by private and semi-public agencies are provided by the Tennessee Valley Authority, the American Gas Association, the Edison Electric Institute, the National Coal Association, the National LP-Gas Association, and others.

Information on public lands and park resources can be found in publications issued by the Bureau of Land Management, the Department of Agriculture, the Department of Interior, the National Park Service, the recently disbanded Heritage Conservation and Recreation Service, and a number of similar agencies within the Department of Commerce and Department of Interior. Due to current changes and reorganizations in federal agencies, the exact location and title of publications and agencies issuing them is almost certain to change dramatically. The kinds of information published will remain largely unchanged, however, and the only difficulty is locating the appropriate reports under whatever newly defined governmental agency is currently responsible for them.

INDIVIDUAL STATE SOURCES

Information similar to that described above is often published by state agencies corresponding roughly to the federal agencies mentioned. Considerable variability exists among states, however, and changes in administration also result in realignments of data collection and publication responsibilities among state agencies. A few illustrations of the kinds of information available at the state level can provide an idea of what a search of local libraries and state information offices might yield.

Most states issue some form of annual agricultural statistics, which might include information on forests, fisheries, and wildlife depending on whether such activities are located within the department of agriculture or in some broader agency like a department of natural resources. Similarly information on minerals, mines, natural gas, and oil is usually available but may be published by either a department of natural resources (possibly even a state department of energy) or a department of mines.

Most states have some kind of department of education, which issues annual statistics on public education, school enrollments, state aid to education by local jurisdictions, teachers, salaries, and so on. In some states, these departments also issue annual reports on college and university activities.

A department of labor, possibly in conjunction with a department dealing with employment services, usually issues monthly and annual reports and projections on labor force characteristics and employment levels for local areas and for specific industrial groups. The state highway department or department of transportation usually issues monthly and annual statistics on travel characteristics, gasoline sales, miles of highways built and maintained, condition and traffic of railroads, volume of traffic in airports, and similar categories.

A state department of social services or welfare provides annual reports on numbers and economic condition of the population, together with information about the kinds and levels of public assistance provided. State health departments provide information on births, deaths, diseases, marriages and divorces, and sometimes on levels of hospitalization, mental health levels and treatments, and so on. Frequently the state health department is the source of current population estimates and population projections for the state and localities within the states. Similar agencies often provide data on numbers of doctors, nurses, and other health professionals practicing in local areas within the state.

Information on vehicle registrations, traffic accidents, drivers' licenses, and so on are usually provided in reports by either the department of state or possibly in reports by the state police department. Information on state and local government organization and financial levels is usually maintained and published by a state-level department of state or by a bureau or agency involved in municipal assistance, planning coordination, or similar type of information and advisory activities. A state housing authority may provide information on housing stock, condition, and new construction.

LOCAL INFORMATION SOURCES

Each locality keeps many records for both private and public purposes. While many of these records are technically open to the public, in practice they are frequently inaccessible because of the way in which they are kept and stored. Ledger books, 3 X 5 cards, old cartons or footlockers filled with file folders, and so on are often the findings of planners seeking information on even fairly recent local events. Newer methods of recording and filing information, including computer cards and tapes, are beginning to become more common, but such practices are usually later in arriving in local offices than in state or national offices or in universities. (Chapter 7 provides some information on these newer methods of storage and retrieval.)

Although information kept by local business firms is usually better maintained, it is often more difficult to obtain access to it due to the concern of private business with protecting company secrets, as well as preserving their customers' privacy. Careful approaches plus clear and unimpeachable guarantees of anonymity of individual records are usually necessary before such information will be made available. A number of sources of local information might prove useful for planning purposes.

Tax and assessment records in local governmental offices: These usually provide data on ownership, size and type of building, property uses, assessed value, condition, recent improvements or additions, a record of recent sales and sometimes photographs of the property in question.

Building inspector's records: These are usually combined with tax and assessment records but may be more current, giving the characteristics and cost of recent improvements, results of any recent inspections, and possibly current information on occupancy characteristics of buildings.

Local real estate board: These agencies usually maintain extensive files on properties recently put up for sale. Occasionally the records are maintained over many years, providing a handy historical source of sales information, as well as detailed descriptions of most of the locality's buildings and properties. Especially useful when working in rural areas is a county or township plat book, often published at low cost by such organizations to assist in real estate sales. Such a book, taken from public ownership and plat records, provides rather detailed maps and descriptions of the ownership of larger land segments throughout the local county or township.

Police or sheriff's office: Such offices usually maintain elaborate records of complaints and crimes by place of occurrence, residence of the offender, and type of crime. These records are sometimes difficult to gain access to due to practices protecting each individual's privacy. But a cooperative arrangement to obtain such information for small geographic areas may be possible with some careful preparation. Information in these records may provide useful indexes of problem areas in the city within which some kinds of planning activity may be more or less appropriate. For example, a concentration of residences of offenders in one area and occurrences of the same offenses in a different area may reveal a commuting pattern of local offenders, which should be taken into consideration by both local law enforcement agencies and local planning agencies.

Fire department and fire insurance records: Considerable information is usually maintained by most fire departments about the size, layout, and condition of most buildings, especially larger commercial and residential structures. These data are frequently rather recent due to periodic fire inspections and may prove valuable in preparing preliminary surveys and evaluations of certain sections of the city. But many fire departments are hesitant to release such information to others since such disclosure might make future inspections and cooperation much more difficult. Fire insurance reports and ratings often reflect a combination of information on building condition and use with availability and quality of fire prevention and firefighting equipment. Combined with other more detailed information on land use, such data might provide a useful evaluative device for areas of the city or a relatively independent cross-check of the quality of other information. Data about the condition of buildings in the potential redevelopment area of Middlesville and the adequacy of existing equipment to provide appropriate protection for the proposed new developments might be useful for either side in the public controversy over gentrification and redevelopment in the downtown area.

Water and sewer departments: Often combined in a single department, these agencies have records of the number of meters installed and volume of water consumed by individual house or address by month or quarter. While obtaining such information for individual households or businesses is unlikely, arrangements could probably be made to obtain aggregate data on water usage by district or neighborhood of the city. Since such aggregated data are of limited utility to most water and sewer departments, the planner must expect to pay for or provide most of the effort required to obtain the information. In some instances, the city may be considering installing meters on an area basis that could provide useful information for specific sections of the city. If so, some effort to try to make water and sewer districts congruent with local neighborhood planning districts is worthwhile since it may yield a useful source of information on neighborhood activities in the future.

Water usage is surprisingly stable and can give a good indication of the population of an area for which usage figures are known. Differences in social values and life-styles are reflected in watering lawns and gardens or in washing cars, however. Such variations produce significant seasonal differences in consumption patterns in different parts of the city related to social class and must be taken into account in making population estimations.

Electricity and gas companies: These companies are usually privately owned and operated. Although they have information on the number of meters installed and the usage rates for individual households and other buildings, most companies are reluctant to release information to other companies or public agencies. A carefully prepared approach is called for, with guarantees of nondisclosure of information on individual houses or units. Providing the background work for tabulating their data in larger geographic areas will almost certainly be necessary. As with water usage, consumption of electricity and gas is a relatively constant ratio to number of persons or type of activity, but seasonal and socioeconomic variations must be controlled for.

Telephone and usage: The local telephone company has records of the number of telephones installed locally and their volume of usage. The company is usually reluctant to make such information available to local authorities, frequently citing state and federal regulations limiting the use and distribution of such information. The key to acquiring the data probably lies in full guarantees of protection against disclosure of individual information and providing all or most of the work required to tabulate the data by larger geographic areas. While number of telephones and usage data are usually less reliable as indicators of numbers of persons present, they may be more useful in some rural and less developed areas where gas, water, and sewer services are not available.

Sales tax receipts: In states and localities that collect sales taxes, information on receipts is frequently reported by local political jurisdiction. Total receipts as well as receipts for individual types of commodity may be a useful indicator of total population size, as well as the economic condition of the population. Receipts for basic necessities such as food or clothing are generally more reliable indicators, but these commodities are often exempted from sales taxes to minimize the impact on lower-income families. When using sales taxes as an indicator of changes over time, pay particular attention to both the effects of inflation on increasing the volume of receipts, as well as to changes in the tax laws regarding what kinds of commodities are to be taxed and the tax rate.

County health office: The local county health office is the principal data-collection and referral agency for almost all information about births, deaths, diseases, accidents, and a variety of other vital statistics, which are ultimately reported by state and federal agencies. Some assistance in handling the basic records of the health office could establish the basis for a useful long-term cooperative arrangement between local planning and health offices that share a number of interests and problems.

County extension programs and the soil conservation service: Although these activities focus primarily on rural and farming issues, they share a number of interests and information needs with local planning agencies, particularly with county planning agencies. These agencies usually have detailed records of the ownership and type of crops raised of most local farms and also detailed information on soils, slope, drainage, wildlife, farmponds, and hunting areas. Of particular interest to local planners may be their usually well maintained supply of aerial photographs and soil surveys of local areas. Such agencies often have copies of a number of the publicly available earth satellite photos of local areas, some taken with infrared and other advanced techniques.

Local school boards: Most local school boards maintain some form of school census, often required by state law, which provides a basis for anticipating future demands for educational facilities by determining the number of young children living in individual school districts. Conducted by teachers or members of the local Parent-Teachers Association, such censuses are often quite accurate, though highly limited, counts of local population. Adults are generally ignored except as the parents of children, and childless households tend to be dropped from the counting process quite early. With some advance preparation and cooperation, significant improvements in the school census operation and in its utility to the local planning office are possible. School administrators are usually willing to release these data on a district-by-district basis, although individual household information is usually not released. School enrollment data can be useful to local planning offices in making population estimates and estimating local migration patterns. These data are usually easily available from local school districts or boards of education.

Automobile registration and drivers' licenses: These records are usually maintained by state offices, but information for local areas can often be obtained. Both registration statistics and the number of driver licenses issued provide information about the numbers of local drivers, as well as age and sex information and some data on types of automobile ownership. The cooperation of state agencies in providing such data to local offices is problematic and is often best handled through state political party channels.

Local advertisements and want ads: Frequently overlooked as a source of information are commercial advertisements in newspapers and in the Yellow Pages of local telephone directories. Although obviously biased toward only those establishments that choose to pay for advertisements, such information can be used to provide a relatively sensitive picture of local activities and short-term trends. Some initial calibration is needed by noting the type and volume of advertisements occurring during some period when the actual volume and type of business activity is known from some other source. Calibrating during census years, particularly economic censuses, is the obvious solution. Once these first benchmarks are established, short-term and long-term trends can be charted in a number of areas of activity. Possibilities include housing vacancies, jobs and unemployment, rental and sales values for houses and apartments, and number of doctors, dentists, and nursing homes. Carefully constructed and maintained over a number of years, such indicators can prove useful and sensitive barometers of local activities. Local residents in Middlesville trying to document the need for

maintaining lower-priced housing instead of providing more luxury apartments might find a compilation of low-, medium-, and high-cost rental advertisements useful in their public presentation.

Local social service agencies: The number of applicants and payments to local population for welfare, Aid to Dependent Children, unemployment compensation, and so on can provide useful information on the state of the local economy and its short-term impact on lower-income persons. These data are available only in aggregate form, and some prior arrangements with local offices to aggregate statistics in the appropriate form are necessary. Even without statistical information, systematic observation of the number of persons in line at such agencies is a simple and sensitive indicator of short-term trends in the local economy.

League of Women Voters and chamber of commerce: Both of these groups collect and maintain elaborate files and data sets on local activities for their particular constituencies (local voters in the former case and local businesses in the latter instance). In many situations these files are of high quality and can be of considerable assistance to the local planning office. Reasonable access is usually possible since the information is intended for the public in both cases. The names of local public officials, budgets, minutes of public meetings, lists of local businesses by type, estimates of local sales volumes, and so on are the kinds of data that might be found in these organizations. Both types of organizations frequently prepare special reports on specific local issues and carry file copies of special studies conducted by their state and national affiliates.

Local newspapers, magazines, and libraries: Such organizations usually maintain extensive files and archives on local happenings. These often include newspaper clippings, photographs, minutes of meetings, newsletters, and correspondence. Similar archives and files may be found in local planning offices together with copies of old plans, proposals, and working documents on zoning hearings and subdivision reviews.

Economic development districts, councils of government, and similar agencies: A number of regional agencies are to be found in most localities, usually combining the data and concerns of a number of local governments and relying heavily on federal sponsorship to establish and maintain their local legitimacy. The utility and the existence of such agencies fluctuate tremendously from region to region, as well as with whoever is in control of the national presidency and Congress. Usually one or more such agencies are to be found locally, however, and they can provide additional sources of information for local planning officials. Frequently such data are merely compilations of data collected from more localized authorities, but sometimes they are reported more clearly and consistently than the same data maintained in the local offices from which they originated. These regional agencies also tend to provide a ready basis for comparing local performance of a number of communities at once. And they may have better and more consistent access to state and federal data sources than the local office or library can provide. Their files and library services are worth becoming familiar with as a potential adjunct to information maintained in local planning offices.

CONCLUSION

The information provided in this chapter only begins to scratch the surface of the kinds of data sources available to knowledgeable and imaginative local planners. A few hours in the library can save weeks of effort and thousands of dollars of costs once some knowledge of how, where, and what is available has been acquired. Indeed many so-called experts on various topics are knowledgeable only about how to find information from diverse sources. Except for this valuable skill, they often know nothing more than the average person. The information offered here should provide a useful springboard to developing your own skills over time.

APPLICATIONS

1. Consulting a recent copy of the *County and City Data Book*, select a nearby city of about the size of Middlesville (60,000). Find its population for each decade from 1940 to 1980. Compute its average annual rate of growth during each decade. (See Chapter 4 on analytical methods.)
2. Find the census tract and block statistics volumes for the city selected. For each tract, record the total number of houses and the number with inadequate plumbing. Record and compare the total population of city blocks on the eastern boundary of the city with the total population of city blocks on the western boundary of the city.
3. Using the appropriate volumes of *Current Population Reports*, determine the estimated population of the city for 1975 and the current year. Evaluate the probable accuracy of these estimates based on your knowledge of growth rates during the 1970-1980 period and your own knowledge of local events since the 1980 census.
4. Visit the county health department and secure information on the number of births in the city from March 31, 1975, to March 31, 1980. Compare these births with the total number of persons aged 0 to 4.9 years reported in the 1980 census. How would you account for any discrepancy between the two figures? Does this tell you anything about migration to or from the city over the 1975-1980 period?
5. Attempt to get information from the local water department on water consumption for the 1970-1980 period. Compute the number of gallons per person per year for 1970 and 1980. Compute the average annual rate of change in total water consumption from 1970 to 1980. Compare this figure to the average annual growth rate of population for the 1970-1980 period computed above.
6. Visit the local chamber of commerce and determine what information it can provide on the number, type, and size of local businesses over the past five years.
7. Visit the local school administration headquarters and determine what information it has on school enrollments and school censuses, if any, over the past five years. Compare school enrollment and school census figures for 1980 to the census results for 1980. How reliable would school enrollment or school census figures be as an estimator of total population between censuses?

PART II

INFORMATION ORGANIZATION

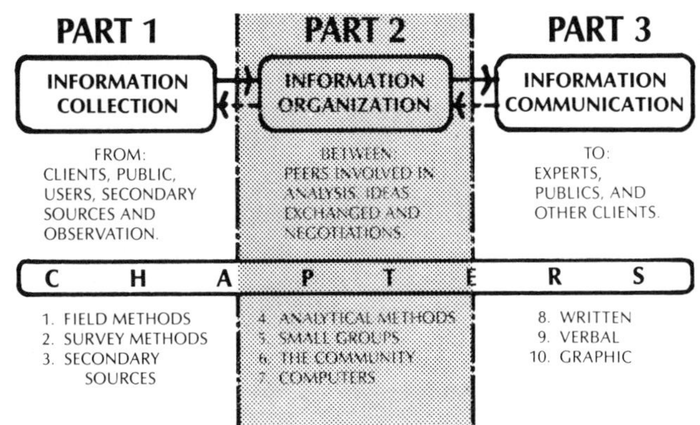

Collected information must be analyzed, processed, and understood before it is useful to planners. The chapters in this part describe how planners organize information so that it makes sense and is professionally useful. They consider analytical methods, their use in certain contexts, and new and evolving techniques in the area. Chapter 4 describes analytical methods employed in planning. Chapters 5 and 6 deal with process. Chapter 5 covers the interpersonal skills and techniques of working effectively in small groups and Chapter 6, aspects of involving the community in planning. Chapter 7 describes the role computers can play in information gathering, analysis, and communication.

4

Analytical Methods

Allan Feldt and Mitchell Rycus

Information cannot usually be disseminated in the same form in which it is collected. Data and observations must be organized and processed to make them understandable to both the planner and to the audiences for whom they are intended. Important findings must be separated from unimportant ones and presented in a clear, convincing, and understandable manner.

A number of highly sophisticated and powerful analytic methods are available, but for most planning purposes only a few elementary techniques are necessary and appropriate for preparing information for dissemination. The quality of the data, the sophistication of the audience to be addressed, and the abilities of the person organizing the data all combine to dictate a fairly simple and commonsense approach to analyzing and digesting most information prior to attempting to communicate it.

Finding and choosing the proper technique for a particular communication requires a careful balancing of sophisticated background and simplicity of approach. This chapter presents a broad array of many of the simpler techniques appropriate to planning problems. Some knowledge of elementary statistics or mathematics is necessary for some of the techniques described, but an understanding of the analytic choices available and how they match the requirements of the communication problem being faced is the most critical ingredient for the successful organization and presentation of information.

At the simplest analytical level, communication consists of descriptions using numbers and percentages. How many vacant houses, how many houses in deteriorated condition, what proportion of all houses each of the groups consists of, and where they are located are all questions likely to arise in a neighborhood redevelopment situation such as that described for Middlesville. Simple measures of distribution and composition are usually adequate for most such analyses.

At a slightly more sophisticated level, it may be desirable to know how many deteriorated houses are vacant or occupied. A two-way table presenting the appropriate cross-tabulation is called for. But before such a table is presented to an audience, it is sometimes desirable to be certain that the apparent relationship between housing vacancy and deterioration is real—that is, that it is not merely one that could have occurred by chance. This calls for a test of statistical significance, possibly accompanied by some measure of association describing the strength of the relationship. Several tests and measures of association are described in this chapter.

A multiple cross-tabulation may sometimes be necessary in order to eliminate the effect of a third attribute on the relationship being examined. This might occur in Middlesville if it were suspected that older houses differed from newer houses in level of deterioration regardless of vacancy status. On rare occasions it might be appropriate to examine and demonstrate the interrelationships of three or more variables with one of several types of multivariate analyses. Such techniques, however, are usually well beyond the quality of the data available, as well as beyond the understanding and appreciation of the audiences for whom the information is being organized.

Organizing complex decisions, anticipating future conditions, and keeping track of the current state of the system as it changes over time are all activities necessary to the local planning office. Communicating dynamic forms of information such as these is especially difficult. A number of techniques are presented to help solve this kind of presentation problem.

Finally it is sometimes desirable to demonstrate how a particular process operates. Although a number of highly sophisticated mathematical modeling techniques are available for such purposes, several non-quantitative techniques such as simulation/gaming and scenario writing may be more appropriate to the communication problems being considered here.

STATISTICAL METHODS

One of the most useful methods of numerical analysis available to planners is statistics. In all likelihood planners have learned some statistical techniques before entering the profession, but a review here of some broad areas of statistical methods may be useful.

First, there are some problems associated with the indiscriminate use of statistics. The presentation of a numerical analysis seems to imply that some absolute truth has been presented. Numbers are something that most people can agree upon, and they lend themselves to making normative judgments on various issues. On the other hand, there is Mark Twain's assumption that "there are lies, damn lies and statistics." But such a cynic probably would not accuse a physicist who used quantum-statistical mechanics to describe certain atomic behavior of lying, even though the statistical methods are essentially the same. In fact, by either sloppy analysis or improperly using certain statistical methods, some erroneous conclusions have been stated. The planner or social scientist who does an improper analysis causes the public to realize the fallaciousness of his or her arguments; hence the perception that "numbers can lie."

Complex statistical analyses should be done by competent statisticians. Planners who are not competent in these areas should at least be knowledgeable about what statistics can and cannot do and when or when not to use them.

Measures of Central Tendency

The most frequent analyses are measures of central tendency. Such terms as *average, mean, mode, median* and the like are used to describe the most prevalent value or values of various population characteristics. The average price of a garbage pail, the median family income, and the modes of vehicle traffic volume are typical of the measures planners may be concerned with. Indeed, along with percentages, central measures are the most common numerical analyses planners use. But stating these measures without some additional qualifying information could prove harmful. To state that the average price for four garbage pails is $30 and not state that three are around $5 and one is $100 can be misleading. One must also state some measure of this variation or at least the range of the values.

Dispersion

The term *dispersion* is used to describe the measures of variation in the data. The more common terms used are *variance, standard deviation,* and *range.* To say that the average price of a garbage pail is $30 and the range is from $5 to $100 is a better representation of the data than just stating the average. One at least sees that the average is closer to $5 than to $100.

Another example is to say that the average income of a group is $20,000 per year with a standard deviation of $4,000 per year. This is also more meaningful than stating just the average. Indeed not mentioning a measure of dispersion when presenting an average could easily be misleading and may even be dishonest. In the average-income example one might claim that well over 90 percent of the population being considered has incomes between $8,000 and $32,000 per year (the average plus or minus three standard deviations). Suppose, however, that the standard deviation was only $2,000 per year; then the range would be between $14,000 and $26,000, a much more homogeneous income group. Now if one were trying to gain support for an income-assistance program for families with incomes less than $10,000 per year, a considerably greater amount of funds would be needed for the group with a standard deviation of $4,000 per year than for the group with a standard deviation of only $2,000 per year even though both groups have the same average income. Simply stating the average income value in this case would not be sufficient for decisions concerning an income-assistance program. Some measure of dispersion must also be presented. But even this—an average and standard deviation—is not always a complete picture of the problem.

Distributions

It is important to understand how the group is distributed among the various levels of the characteristics that one is interested in. Describing the number of families in income categories of $0 and $5,000, $10,000 and $15,000 per year, and so on presents a more complete picture. Plotting this information with the number or percentage of the individual listed on the side and on the bottom listing the intervals of the measure provides an even better understanding. Such plots, called *histograms*, are often revealing and can assist the decision maker even more than just the central measure and dispersion values.

In the example of the income groups, suppose the two plots shown in Figure 4.1 represent the actual data for two groups *A* and *B*. These graphs show that the average and standard deviations are not sufficient to indicate the differences between the two groups. Clearly the group with a smaller standard deviation has a larger percentage of individuals with income under $10,000 per year than does the other group. Showing the actual distribution would be vital in this case since obviously the groups are not normally distributed as one might have inferred from the average and standard deviation.

Fortunately most large populations are normally distributed (the famous bell-shaped distribution) or distributed in ways that can be expressed mathematically. And it is this knowledge of population distributions, with some understanding of probabilistic events, that allows the statistician to do the more sophisticated analyses. The discussion to this point has been about concepts that should be familiar to most planners, but the discussion that follows is about techniques best left to planners skilled in statistics.

Test of Significance

The probabilistic nature of inferential statistics requires a knowledge of the frequency with which events can be expected to occur. Most people are

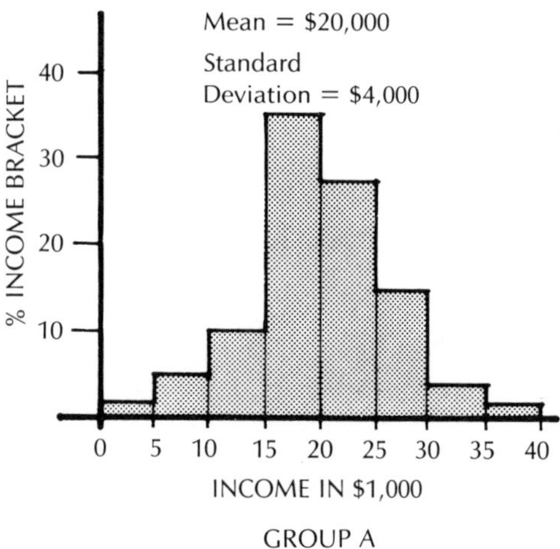

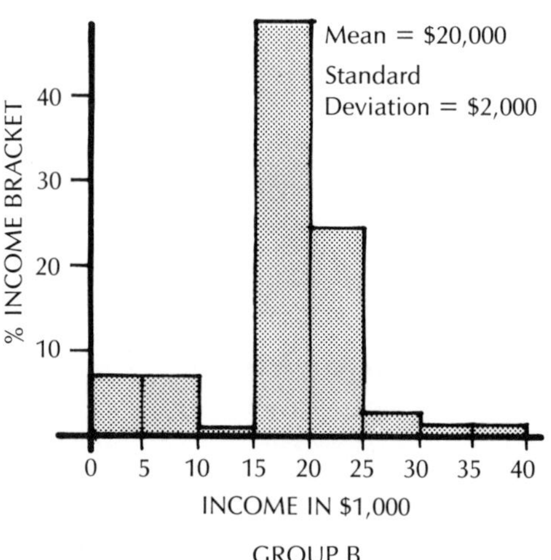

Figure 4.1 Income distribution plots.

aware that a large number of coin flips should produce as many heads as tails. We say that the probability of getting a head or tail on any given flip is 0.5. A percentage can also be expressed as a measure of probability. Figure 4.1*A* shows that the probability of a family's having an income under $10,000 per year is about 0.07. If the population were normally distributed or distributed in some well-defined manner, then certain statements about the percentages in each interval could be made and tested to within some defined probability without knowing the actual distribution.

Such hypothesis testing is done quite frequently. A small sample from a larger population can be taken (as described in Chapter 2 on surveys) and analyses conducted to make inferences about the larger population. When one considers how large the size of the populations are that planners usually deal with, one appreciates how statistical analysis can enable one to make inferences quite accurately about the larger population using a relatively small number of data. Estimates of population characteristics, such as average annual income, or the average number of rooms in a residence, or even the average stream flow in a river could not be made accurately without the benefit of inferential statistics. The level of significance, which is the probability that an error is made in either accepting or rejecting valid hypotheses, can be stated for the user's assessment of the hypothesis.

Statistical analyses involving probabilistic assumptions about one population or for making comparisons between more than one population are common. Most analyses are used either to accept or to rule out relationships that are not always apparent between certain events occurring in the population: Is there a relation between cancer and industrialization? Is poverty related to educational level? Statistical methods cannot determine causal relations but simply give a measure of probability associated between the occurrence of various events. One may find that a statistically significant relation exists between cancer and industrialization, but that relation does not imply that industrialization causes cancer. The real value of the analysis is in allowing the planner to predict possible future events based upon significant statistical relations, using the occurrence of one event to predict the occurrence of another.

Predictive Models

Sometimes one is able to show that a relationship exists between events without being able to say what is behind them. For example, if it can be shown that there is a high correlation between the number of televisions in an area and the number of people developing cancer in that area, then one may be able to predict cancer rates based upon television sales. But banning television sets in the area will not significantly reduce the number of cancers because it is not the sets that cause cancer; it just happens that the two events are highly correlated.

Predictive models of this nature are usually in the form of a regression analysis, where one variable is compared to one or more other variables to determine the strength of association between them. If there is statistical evidence to accept an association, then predictions about future events can be made. For example, determining whether traffic flow is affected by the different types of land use in the area is easily accomplished with a regression analysis. If it can be shown that increases in commercial and industrial uses are related to increased traffic flow, then an estimate of the increase in traffic flow is obtained by stating the potential levels of commercial and industrial land use.

Summary of Statistical Analysis

Many more complex methods involving sophisticated sampling techniques and the use of computers are available to planners. For the most part, planners will perform analyses only at the descriptive level, and only a few will get into the higher levels of analysis such as hypothesis testing, regression analysis, analysis of variance, nonparametric statistics, and the like. But it is important for planners to be aware of the more sophisticated statistical techniques available and, maybe even more important, to know when not to use them. Frequently a complex numerical analysis is presented when a simple analysis is just as effective. For instance, when a regression analysis is performed, a simple curve fitting technique would often be sufficient. Some people will try to impress others with an inordinate amount of numbers and complex equations. But the well-trained planner should know the limitations of these techniques and use them only when appropriate.

POPULATION COMPOSITION AND DISTRIBUTION

A considerable number of analytic techniques can be used to display and analyze the nature of a population; however, six fairly simple techniques and measures provide an adequate basis for many analyses and should be understood by all planners.

Distribution Map

The distribution map is a map of the area in question broken down into geographic subunits such as city blocks or census tracts for which appropriate data are available. Aggregate population characteristics such as median income, average age, median education, percentage of housing units built before 1950, and similar measures are then indicated for each subunit either as numerical data or as various colors or shadings of gray keyed to a map legend. (See Figure 4.2.) Population characteristics are sometimes best portrayed as a percentage of the total population of the subunit, while in other cases the percentages may be more appropriate as percentages of the total population of the area. Percentage values are grouped into convenient categories selected to maximize the clarity of the data being presented. The use of black and white shadings is usually preferred to minimize reproduction costs and to make black and white copies fully usable. (Further information on creating such maps is provided in Chapter 10.)

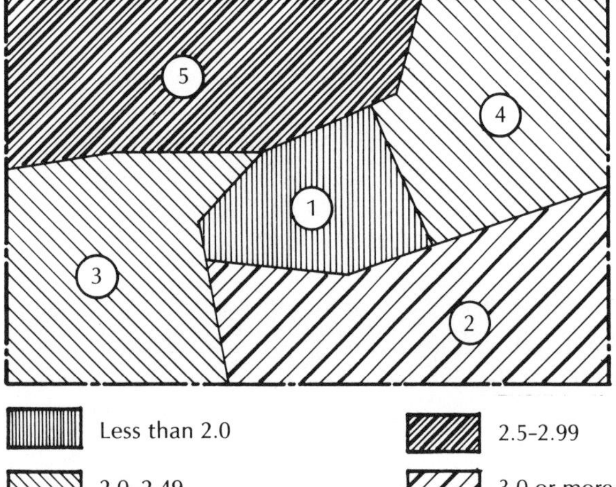

Figure 4.2 Average number of persons per household in Middlesville census tracts, 1980.

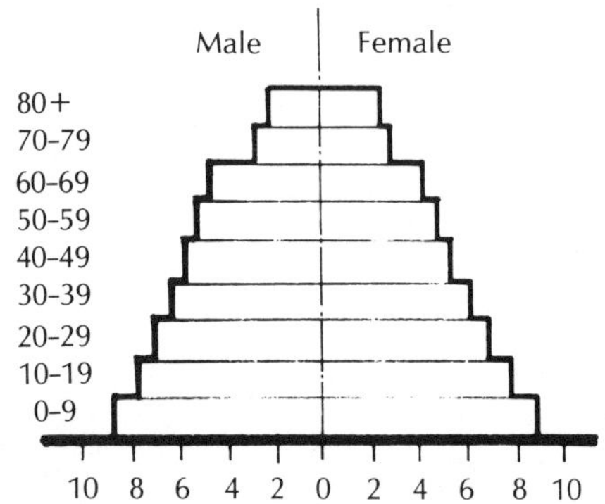

Figure 4.3 Population pyramid of City Opportune.

TABLE 4.1
Population of City Opportune

Age Group	*Males*	*Females*	*Percentage of Total*	*Percentage of Total*
0-9	20,000	19,000	8.8	8.4
10-19	18,000	17,500	8.0	7.7
20-29	16,000	16,000	7.1	7.1
30-39	14,500	14,500	6.4	6.4
40-49	13,000	13,000	5.8	5.8
50-59	11,000	11,500	4.9	5.1
60-69	9,500	10,000	4.2	4.4
70-79	6,000	7,000	2.7	3.1
80 and over	4,000	5,500	1.8	2.4
Total	112,000	114,000	49.7	50.4
	226,000			

Population Pyramid

A population pyramid is a simple but valuable graphic portrayal of the age and sex composition of a population. Pyramids may be computed and displayed for the total population of an area, for important population groups, or for significant subunits such as census tracts or city blocks for which appropriate data are available.

A population pyramid is calculated by converting the number of males and females in major age groups into a percentage of the total population. These percentages are then displayed as a series of bar graphs, with males projecting to the left from a center line, females projecting to the right, and age groups increasing from the bottom bar graph to the top bar graph. Figure 4.3 displays the data from Table 4.1 as a population pyramid.

Population pyramids for two different areas or for the same area at two different time periods may be readily compared indicating shifts in population growth and composition. Full interpretation of such shifts must be based upon the known history of each area or population group, but there are some general guidelines for the kinds of pyramids that occur with certain types of populations. A pyramid such as the one illustrated in Figure 4.3, which shows regular decreases in the percentage of the population from the youngest to the oldest ages and reasonable symmetry between males and females, is often considered to be normal. Such a pyramid presumably reflects a fairly stable population subject to relatively little in-migration or out-migration and with most persons marrying and having two or three surviving children.

A number of interesting and significant deviations from such normal pyramids are found in specific regions, in certain neighborhoods within cities, and even in some national populations, particularly in developing countries. There are four commonly found variations. The first is an area of heavy male concentration, probably reflecting a military base or other institutional population, an area of heavy employment of male workers such as a mining or lumbering area, or a skid row or rooming-house section of a city. The second is an area of heavy female concentration, probably reflecting a similar military or institutional concentration of females or else a local economy employing large numbers of females as clerical, secretarial, or textile mill workers. The third is a typical pyramid found in areas of very high fertility, often in developing countries or in areas where previous high infant mortality has been coupled with high fertility. The fourth type of pyramid is one in which a large percentage of the population is made up of older persons with comparatively small numbers of persons in the middle child-bearing years and relatively few children. Such an area is frequently found in older rural areas subject to heavy out-migration of younger- and middle-aged persons leaving a residual population that is dominated by older persons not willing or able to move to new locations. (See Figure 4.4.)

Coefficient of Dissimilarity

The coefficient of dissimilarity is useful in comparing the distribution of two population properties among a set of geographic sub-areas with each other. The two properties may be either the same characteristic at two points in time or different characteristics at the same point in time. In the former case the coefficient gives an indication of the degree of change or shift in the population that has occurred over time. The measure is sometimes called a *coefficient of redistribution.* In the second case the coefficient gives an indication of the degree to which the two characteristics are equally distributed over the same

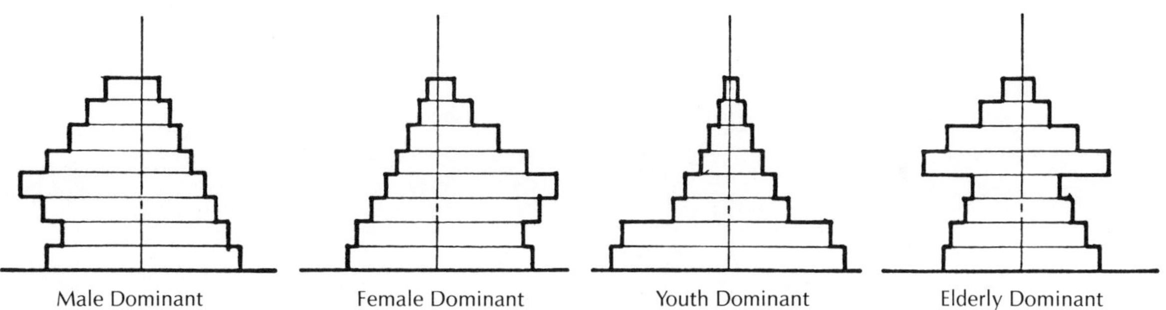

Figure 4.4 Illustrative population pyramids for four different types of population.

areas. The value is sometimes called a *coefficient of separation* or an *index of segregation*. Values of the coefficient of dissimilarity can vary from 0.0 to 100.0, the magnitude of the number being interpreted as the proportion of one of the population properties that would have to be redistributed in order to produce a distribution equal to that of the other population property.

An important qualification of the usefulness of the coefficient is the fact that the value attained depends in part on the number and size of the geographic subunits being analyzed for distributional patterns. Thus an index of segregation between blue-collar and white-collar workers would be larger when city blocks are used as the analysis units than when census tracts are used. Comparing two coefficients of dissimilarity to each other is meaningless unless comparable geographic units are used in the computation of both of them.

Computation of the coefficient of dissimilarity is simple. List the values of population *A* and *B* or characteristics *A* and *B* for each of the geographic subunits being considered. Convert each of these numeric lists to percentages based upon the total value of *A* and total value of *B*, respectively. Subtract the percentage of *B* from the percentage of *A* and list *D* (for "difference") in a separate column, retaining the sign of the remainder. The value of the coefficient of dissimilarity is then the sum of either all the positive *D*s or all the negative *D*s, which, as a check on computations, should be approximately equal. The sum of the absolute values of *D* divided by 2 is an alternative method of computing the coefficient of dissimilarity, but it is less desirable since it does not provide the automatic computation check of comparing the positive and negative sums of *D*.

A comparison of the degree of separation of white-collar and blue-collar workers in City Opportune in 1970 and 1980 could be made to determine whether social separation of the population was increasing or decreasing over time. Assuming for simplicity that Opportune contains only five census tracts or other relevant geographic subunits, the computation and comparison would be as shown in Table 4.2. The example indicates that there has been an increase in separation between white- and blue-collar workers in City Opportune over the decade due in large part to the substantial increase in white-collar workers over this period, some of whom settled in tracts 1, 2, and 3, but most of whom settled in tracts 4 and 5. The coefficient of dissimilarity increased from 25.0 to 33.4, reflecting the magnitude of this increase in social separation.

Similar analyses can be constructed showing changes in income classes, racial or ethnic groups, major types of land uses, and so on. The coefficient of dissimilarity is relatively easily computed and displayed, but its dependence on the geographic units of analysis employed must be considered in each application.

TABLE 4.2
Degree of Separation of White-Collar and Blue-Collar Workers in City Opportune, 1970 and 1980

	1970					1980				
	Number		*Percent*		*D*	*Number*		*Percent*		*D*
Tract	*WC*	*BC*	*WC*	*BC*		*WC*	*BC*	*WC*	*BC*	
1	2,000	8,000	10.0	26.7	−16.7%	3,000	9,000	10.0	30.0	−20.0%
2	5,000	5,000	25.0	16.7	+ 8.3	7,000	4,000	23.3	13.3	+10.0
3	4,000	6,000	20.0	20.0	0.0	5,000	5,000	16.7	16.7	0.0
4	3,000	7,000	15.0	23.3	− 8.3	4,000	8,000	13.3	26.7	−13.4
5	6,000	4,000	30.0	13.3	+16.7	11,000	4,000	36.7	13.3	+23.4
Total	20,000	30,000	100.0	100.0	25.0	30,000	30,000	100.0	100.0	33.4

Note: WC = white collar; BC = blue collar.

Lorenz Curve

The Lorenz curve provides a useful graphic illustration of the degree of separation between two population characteristics and is directly related to the coefficient of dissimilarity and the Gini concentration ratio. The same data used for calculation of the coefficient of dissimilarity are employed here in a slightly modified array to develop a Lorenz curve.

First, the census tract data are reordered on the basis of increasing significance of one variable, in this case the increasing preponderance of blue-collar workers over white-collar workers in each census tract. Another variable totally unrelated to the two variables being compared could also be used, such as distance of the tract from the center of the city or median income of each tract. In such a case a more complicated interpretation of the meaning of the curve would have to be developed, corresponding to the interplay of three rather than only two variables. In any event the tracts are reordered on the basis of the significance of the variable selected, and the percentages computed are added cumulatively. These cumulative percentages are now plotted on a simple graph and the resulting curve compared to the diagonal line that could be drawn from one corner of the graph to the opposite corner. The diagonal represents a curve of nonseparation—that is, the plot of cumulative percentages from two equivalent percentage distributions. The computations and curves are illustrated in Table 4.3 and Figure 4.5.

The shaded area between the diagonal line and the curve in Figure 4.5 illustrates the degree of

TABLE 4.3
Cumulative Percentages of White-Collar and Blue-Collar Workers in City Opportune, 1970 and 1980

	1970				*1980*			
	Number		*Cumulative Percent*		*Number*		*Cumulative Percent*	
Tract	*WC*	*BC*	*WC*	*BC*	*WC*	*BC*	*WC*	*BC*
5	6,000	4,000	30.0	13.3	11,000	4,000	36.7	13.3
2	5,000	5,000	55.0	30.0	7,000	4,000	60.0	26.6
3	4,000	6,000	75.0	50.0	5,000	5,000	76.7	43.3
4	3,000	7,000	90.0	73.0	4,000	8,000	90.0	70.0
1	2,000	8,000	100.0	100.0	3,000	9,000	100.0	100.0
Total	20,000	30,000			30,000	30,000		

Note: WC = white collar; BC = blue collar.

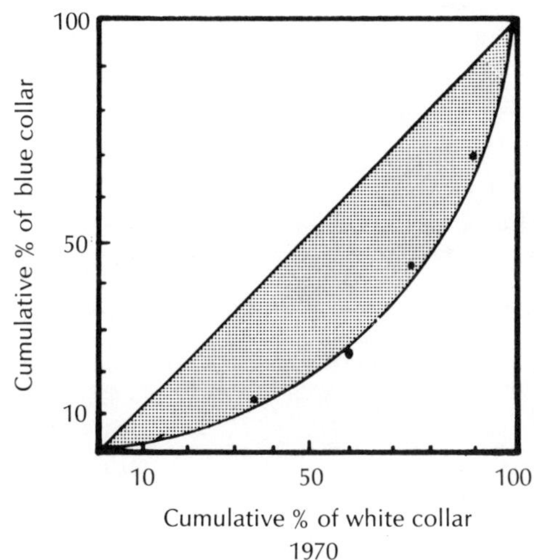

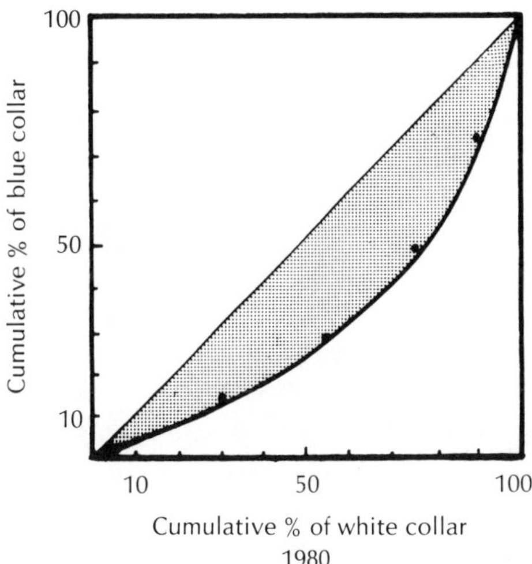

Figure 4.5 Cumulative distribution of white- and blue-collar workers in Opportune census tracts, 1970 and 1980.

separation between the two population characteristics for the geographic units of analysis utilized; the greater the area, the greater the degree of separation. It can be shown that the maximum vertical distance between the diagonal and the curve is equal to the value of the coefficient of dissimilarity. The actual area of the curve can be computed by a formula known as the Gini concentration ratio.

If the census tracts are ordered on a different third variable before cumulating the percentages, the Lorenz curve illustrates the comparative distribution of the first two variables relative to the third ordering variable. Thus if tracts were ordered by distance from the center of the city, the Lorenz curve and its associated measures reflect the centralization of white- and blue-collar separation.

Gini Concentration Ratio

The Gini concentration ratio is a more precise measure of the degree of separation between two population characteristics and is a close approximation to the measurement of the total area between the diagonal and the curve of a Lorenz curve. The formula for its computation is as follows:

$$G_i = \frac{(X_{i-1}Y) - (X_i Y_{i-1})}{10{,}000}$$

$$G_{1970} = \frac{3352.5}{10{,}000} = 0.335$$

$$G_{1980} = \frac{4208}{10{,}000} = 0.421$$

It can be shown that the value of G is always equal to or greater than the value of the coefficient of dissimilarity and that it is always less than twice the value of the coefficient of dissimilarity.

Location Quotient

The location quotient reflects the degree of specialization of a geographic unit in a given type of activity where specialization is said to occur whenever an area has proportionally more of an activity than some other reference area. The usual reference area is the larger city, region, state, or nation within which the geographic subunit occurs. The degree of specialization in white-collar workers of the five City Opportune tracts is reflected by a location quotient computed by dividing the percentage of white-collar workers for each tract by the percentage of white-collar workers for the city as a whole. From Table 4.3 we know that in 1970, 40 percent of the workers in the city were white-collar workers (20,000/50,000). In 1980 the percentage of white-collar workers rose to 50 percent of all workers. The percentage of workers in each tract who are white-collar workers and the corresponding location quotient based on dividing by the appropriate total city percentages are given in Table 4.4.

Values of the location quotient above 1.0 indicate specialization in the activity being considered, the amount greater than 1.0 reflecting the degree of specialization. In the City Opportune illustration, tracts 2 and 5 provide concentrations of white-collar workers, with tract 5 being the most specialized in the city. The

TABLE 4.4
Location Quotients of White-Collar Workers in Census Tracts in City Opportune, 1970 and 1980

	1970		*1980*	
Tract	*Percent of Total*	*Location Quotient*	*Percent of Total*	*Location Quotient*
1	20.0	0.50	25.0	0.50
2	50.0	1.25	63.6	1.27
3	40.0	1.00	50.0	1.00
4	30.0	0.75	33.3	0.67
5	60.0	1.50	73.3	1.47

location quotient is used extensively in economic base analyses.

ECONOMIC BASE AND ECONOMIC MULTIPLIERS

It is sometimes desirable to know what industries are the major economic contributors to the growth of the city. Such industries are called basic industries or export industries in contrast to service industries or local consumption industries. Examples of basic industries include many types of manufacturing, most of the extractive industries, and many national or regional service industries such as large universities or hospitals. Examples of service industries include food stores, beauty parlors, restaurants, and schools, hospitals, and universities serving local residents. Jobs may occur in either basic or service industries, but the location of new services generally follows the location of new jobs, ultimately in basic industries rather than in service industries.

Identifying and using information on the economic base of a city poses a number of problems, the most important of which revolve around the definition of an industry. Data on industrial employment and activity from the U.S. Census Bureau are usually employed, and these define industries according to the Standard Industrial Classification system, commonly known as the SIC code. For purposes such as defining an economic base, these classifications are fairly crude and cloud over the fact that most businesses are involved in a number of different activities, some of which are basic but some of which are service. In addition to these problems, the gradual shift of employment into more service industries over the past few decades has made even the concept of basic industries increasingly questionable. Nonetheless, projecting the future growth of a city requires that planners take into account the probable growth or decline of jobs in the area since in most cases it is jobs that attract and hold residents to a city.

Two concepts and related measures are employed to address these issues and to allow some crude approximation to understanding and predicting future growth of the city: the location quotient and the economic multiplier. The location quotient measures the degree of specialization of an area in a given activity, in this case an economic activity. If reasonably well-defined basic industries can be identified by exhibiting location quotients well above 1.0 and the future growth or decline of such industries can be estimated, the number of jobs in basic industries likely to be gained or lost by the community in the future can be estimated. Anticipated new jobs can then be translated into anticipated new population, including dependents of workers in basic industries, plus both workers and dependents in the additional service industries likely to arise with the influx of new population. A number of different methods of determining such measures have been developed, more detailed texts should be consulted for a full discussion of the advantages and disadvantages of various methods. The major elements of any method, however, follow a few simple and commonsense steps of development.

1. Examine all the major industries of the city and select those that are logical possibilities as basic industries.

2. Compute location quotients for these industries and select the five to ten industries with values substantially greater than 1.0. In choosing the basic industries, bear in mind that a highly specialized industry employing very few may not contribute greatly to the economic growth of the city even though the value of the location quotient may be high. Some weighting of the number of employees must be given in selecting those industries likely to have a significant impact on future change. Thus a large industry employing thousands of workers with a location quotient of only 1.2 is more basic than a small industry employing thirty to forty people with a location quotient of 5.0.

3. Examine national studies and reports on the

growth potential of the industries identified as basic. Discussions with local plant managers and officials may be useful as well. A definite future time frame such as five, ten, or twenty years must be considered in order to avoid the inherent vagueness of discussions about some possible future ranging from tomorrow to the next century. From this review, develop estimates of the likely percentage increase or decline in the basic industries of your community and convert these percentages into increases or decreases in the number of jobs in basic industries.

4. Determine the multiplier effect by estimating the total number of persons in the community ultimately dependent upon each worker in the basic industries. If all basic industries were clearly identified and all remaining population were made up exclusively of workers in service industries or dependents of workers, this computation would be simple; the total population of the city divided by the total workers in basic industries would equal the economic multiplier. Unfortunately all basic industries have not been identified, all remaining industries are not exclusively service industries, and all population is not dependent upon workers in either basic or service industries. Furthermore, conservative practice in selecting and estimating basic employment leads to non-conservative practice in computing the multiplier since a smaller number in the denominator increases the size of the resulting quotient. Considerable care and judgment in computing a multiplier are called for, and local knowledge is much more important than a series of well-defined formulas from outside experts. A reasonable procedure is to divide total population by total employment in basic industries and then to reduce this figure substantially according to personal judgment and local knowledge, recognizing that the initial figure attained is always too large. A normal range for economic multipliers is generally 4.0 to 8.0.

5. Multiply the number of additional employees expected in the basic industries by the economic multiplier to produce a crude estimate of population growth for the time period earlier specified. Future plans for the city may now take this potential growth into account, including needs specific to the basic industries themselves. The highly judgmental character of the analysis should be kept in mind at all times, and the figures derived should be used only as a rough benchmark for projecting possible future growth for the city.

The same techniques may be employed in reverse to make plans for the decline of a city accompanying the decline or movement of some of its basic industries. Planning for population losses and community decline may be even more important than planning for growth since proper use of resources and capital plant becomes much more critical under these depressed circumstances.

ESTIMATES OF CURRENT POPULATION

The term *population estimate* is generally reserved for referring to an estimate of the current or past population whereas the term *population projection* is used to indicate an attempt to specify what the population will be at some future date. With a little care and some advance preparation, planners should be able to provide reasonably accurate estimates of the current population, probably within 5 to 10 percent of the true figures, at least within the first five years of each official census. Population projections, on the other hand, are notoriously inaccurate regardless of the method used or the amount of effort expended. Errors of 10 to 20 percent within five years are commonplace, and errors of 50 to 100 percent for projections twenty years in the future can be reasonably expected in any situation undergoing fairly rapid change and growth.

Nonetheless most planning offices are frequently asked to provide projections of future populations, and the political necessity of providing such information is unavoidable even though its validity and usefulness are highly questionable. Estimates of the current population are less frequently sought, and many planning offices fail to make estimates of the current population on any consistent basis, although reasonably accurate estimates are fairly easily attained and are a prerequisite for making projections.

Population estimates are most easily accomplished by comparing a known population in the past to the value of some symptomatic indicator, such as automobile registrations, school enrollments, or electric meter installations. The ratio of known population to the level of the symptomatic indicator at one time can then be multiplied by the known level of the indicator at the later time to yield the desired estimate.

The symptomatic indicator chosen should be one for which data are reported annually over the period for which the estimates must be made. All symptomatic indicators tend to be biased toward over-estimating or under-estimating particular age groups or types of population, but these biases can be compensated for by carefully combining indicators to produce estimates either for each appropriate subpopulation or providing several estimates with complementary biases that may be combined into a single more balanced estimate.

Thus the number of children enrolled in school is probably a good indicator of the number of children in a community, particularly for grades 2 to 7 corresponding to ages seven to twelve for which enrollments tend to be most stable. The same indicator could be used to estimate total population, but such an estimate would obviously be biased toward younger families with children in school and would tend to misrepresent both older and younger adults. A separate or a compensating estimate of older adults might be made by using deaths of older adults or number of persons receiving social security payments. Younger adults without children might be better represented by data about new home sales, marriages, new drivers' licenses issued, and so on.

The basic computation would be to determine the appropriate figures at the time of the last census, get the ratio between population and the indicator, and then multiply that ratio by the current value of the indicator. If separate estimates for separate population subgroups are being made, the final estimates should be combined to produce the total population estimate. If separate estimates of the total population are being made, the separate total estimates should be averaged to produce a less biased total estimate. Weighted averages may be employed if some reasonable basis for the weighting of different estimates can be determined. It is usually preferable to prepare separate estimates for significant population subgroups from appropriate symptomatic data since this provides some basis for estimating the age composition of the estimated population as well as the total figures. Lack of adequate symptomatic data frequently precludes this refinement, however, and reasonably good estimates can be expected of the total population provided that at least three or four complementary sets of symptomatic indicators are employed. Following are possible symptomatic indicators with qualifications about their use:

Number of births: To estimate women fifteen to forty-five or children under ten.
Number of Deaths: To estimate any age group but especially those over fifty.
Number of gas meters, telephones, water meters, and similar devices.
School Enrollment: To estimate school-age children or young families.
Water consumption: Some seasonal variations by socioeconomic class.
Sales tax receipts: Some socioeconomic class bias; subject to legislative modification.
Bank deposits: Sensitive to state of the economy as well as population size.
Number of automobile registrations: Some socioeconomic bias.

Several multiple regression techniques can also be employed in making such estimates, resulting in small improvements in their accuracy. A more complete discussion of more complex estimation methods can be found in several of the sources cited in the bibliography for this chapter.

PROJECTIONS OF FUTURE POPULATIONS

A considerable number of different population projection techniques are available, but none has been shown to provide any substantial increase in accuracy over any of the others. It is recommended that making population projections should be avoided whenever possible. But when the issue cannot be avoided, the method chosen should be reasonably simple and cheap since expending great efforts and funds is unlikely to improve the accuracy of the projection. Of the four general types of projection methods discussed here, the fourth is recommended as the most informative as well as one of the most economical.

Cohort Survival Method

The cohort survival method is the most widely used and accepted method of projection. It is also the most costly and possibly most complex method employed. Although it is not necessarily more accurate, the results often dazzle clients and public officials, which probably accounts for the popularity of the method. In brief the projector makes a series of assumptions about future birth, death, and migration rates for the population of the area in question. The known population of the area at the beginning of the projection period is broken down into five-year age and sex groups (cohorts). The assumed death rates are then applied to each cohort as appropriate to produce a new cohort five years older plus an artificially created cohort of new babies aged 0 to 4.9 years. Assumptions about migration are then taken into account and appropriate numbers of persons added to or subtracted from each cohort. The entire process is then repeated until, after three or four iterations, the projector has artificial women descended from the first cohort of artificial babies giving birth to a new generation of artificial babies. Clearly any cohort projection extending beyond twenty years becomes extremely unreliable since a set of simple initial assumptions has been allowed to compound its effects four times.

Although the method is not necessarily any more or less accurate than others commonly employed, it allows fairly refined assumptions about the possible future state of fertility and migration in the area. The method is fairly easily run on even small-scale computers and can produce a large array of alternative projections corresponding to alternative assumptions regarding future trends in fertility and migration. Future trends in mortality are likely to be so stable that most projections do not bother varying these assumptions. The result of a computer run of a cohort survival projection with several varying assumptions is usually a satisfying thick sheaf of papers providing some kind of alternative projection to satisfy everyone and in sufficient detail to mystify almost anyone. Although costly and confusing, the results are probably not a great deal worse than could have been produced by other projection methods.

Curve Fitting

Curve fitting covers a number of possible techniques by which past population growth trends are fitted either to mathematical equations or to simple plots on graph paper and the lines extended into the future for varying periods of time. Numerous alternative curves may be employed both mathematically and graphically, most of which have nothing more to recommend them than that they are able to fit historical data with more or less accuracy. Why fitting past data should be related to fitting future data is a question best left unasked since the answers tend to become more tortuous than enlightened.

If curve fitting is to be employed, it is important to cover a historical period at least as long as the future period being projected.

Ratio Techniques

Ratio techniques rely upon some other authority's projections for some larger area embracing the local city, such as the state or region. The known ratio of the city's population to the larger entity's population at the last census date is then computed and multiplied by the projected population of the larger entity at some future date. Slight refinements in the ratio technique enable the person making the projection to make some assumptions concerning the gradual increase or decrease of the ratio from the earlier time period. This depends on whether the city will play a more or less vigorous role in growth in comparison to the larger entity at the base of the ratio.

Ratio techniques are simple and cheap. Further they make some higher authority responsible for the projections by using their results. In requiring very little thought, however, they are probably among the most widely abused and misinterpreted projection techniques used.

Composite Projection Technique

A composite projection technique combines the advantages of shifting as much responsibility as possible to other authorities with making the ultimate user of the projections aware of the great diversity that occurs in making projections, and thereby cautions him or her against accepting any projection at face value. To make a composite projection, gather up as many other projections for the local area as can be found. Good sources are local, regional, and state planning offices, public health agencies, marketing analyses provided in several national business magazines, occasional projections provided by the Bureau of the Census or the Bureau of Labor, local schools and colleges interested in projecting enrollments, and other similar agencies.

Select a number of the most reasonable of these projections and display them in a single table, listing the populations they have projected for appropriate years. The considerable range in these projections is instructive in itself and should caution clients against putting too much faith in any projections, including those about to be developed. Variations in these figures are made even more dramatic by converting the population data provided into average annual rates of growth for each time interval displayed for each of the projections.

For reasonable accuracy and elegance here, it is wise to use either the compound interest formula ($P_2 = P_1 e^{rt}$) or its close approximation ($P_2 = P_1(1+r)^t$) to compute these growth rates. The average annual rate of change attained by dividing the percentage increase of population over the preceding time period by the number of years that have passed is too crude for use in this context.

An examination of these growth rates usually reveals that some of the projections are quite unrealistic, exhibiting implied growth rates of as much as 5 percent per year. Most cities grow at a rate of 1 to 2 percent per year. Figures substantially greater or smaller than such rates can be considered unrealistic unless unique local circumstances verify such extremely high rates of growth. The *rule of 70* is useful in interpreting annual growth rates. Seventy divided by the average annual rate of growth gives the approximate number of years required for a population to double in size. A growth rate of 5 percent per year means that a city's population would double in fourteen years and a growth rate of 3 percent means doubling in about twenty-three years.

Reviewing and comparing the several projections listed with each other and against known reasonable levels of average annual growth rates usually allows one to select one or another of the projections as more reasonable than the others. If necessary, a new projection may be developed as an intermediate between two existing projections, both of which seem reasonable. Beginning with a base population from the most recent census or a reliable current estimate, annual growth rates may then be inserted into the original formula to produce new projected population figures for appropriate years. Varying rates of growth may be employed in different five- or ten-year intervals corresponding to variations in the assumptions embedded in the other projections being evaluated. If the growth rates employed are reasonably close to those of one of the other projections being evaluated, the proportional age and sex distribution of that projection might be employed to estimate the age and sex distribution of the new projection developed. If the new projection departs substantially from other available ones, however, a full cohort survival projection may be necessary to produce reliable age and sex breakdowns.

This method has the advantage of simplicity and basic integrity since it generates its projection openly in comparison to those already developed by others. Arguments for selecting one or another projection or for developing a new one can be clearly made in the light of the other data considered. The user should be well informed about not only how the specific projection developed was made but also how it compares to other known sources and authorities. Nevertheless no method of projection, including this composite method, produces results likely to be reliable for more than a few years beyond the origin of the date of the projection.

SIMULATION/GAMING

Simulation/games have come into some use since about 1970 as an occasionally useful tool in the planning office. Games have been found to have four major uses, only some of which may directly benefit the planning office.

As teaching devices, games provide a strong motivation to learn. The learning situation is based on experience rather than direct intellectual transfer of information. Many games have been found to work well in the classroom and even more often have been found to be useful supplements to conventional classroom teaching. Selecting the proper game and integrating it well with the classroom material are somewhat difficult, however. Teachers experienced in using games know that they add to the teaching load rather than decrease it because of increased motivation as well as because of the time and effort required to set up and manage a game. While widely used and accepted in schools and colleges, games are probably quite rarely used in most planning offices for teaching purposes.

In enhancing communication between citizens and professionals, games have proven to be particularly useful in many planning offices. Most games do not require high levels of experience or background to play, and the general public can often play such games against traditional experts without suffering too great a disadvantage. Games also serve to disarm opponents in public conflicts since they create a situation within which antagonisms over real issues can be temporarily forgotten. Games have often been used as an icebreaker at the beginning of public hearings and meetings where participants are either modest about participating or so filled with tension over the issues to be discussed that the meeting has difficulty getting underway smoothly. In presenting a game in such a context, the person responsible must expect to look a little foolish and frivolous since the participants usually are very serious and want to get down to business immediately. A well-chosen game with enough time to play it through and a suitable transition from the game to the larger purpose of the meeting can prove very useful in such a situation, however.

Games have proven useful in some research applications by offering a comparatively simple method of gaining an overview of the problem being considered and providing a means for cross-communication among different research approaches. Within this context, games provide a first rough draft of any possible models that may be under development by the research project and make discussion of the emerging model easier among different disciplines and with the lay public. Games can also provide a kind of questionnaire for use among members of the research team and with the general public, inviting persons playing the game to criticize its structure and approach to the problem. By suggesting revisions and additions to the game, participants pool their own knowledge of and insight into the research problem. The use of games in this context is not highly likely within a planning office, although such an application might prove useful as part of a team approach to a large-scale design problem.

The use of games in exploring major policy alternatives has received some attention in a few planning offices and has been widely used in national and international affairs. In this context, a game allows policy makers to explore alternative policies and possible developments deriving from such policies. The result is one or more possible futures, which may then be evaluated in subjective as well as objective terms for their desirability as possible future states of the system under consideration. Such an application could be invaluable in local planning offices, but the games must be selected with considerable care and must bear a meaningful relationship to actual policy alternatives present in the city. Furthermore policy makers themselves must understand the meaning of the exercises to make the best use of the possibilities being presented. At this time, the use of simulation/games in this context in local planning offices is fairly rare. It seems likely that it will be some time before the techniques become widely adopted at this level.

Games may be conveniently grouped into the four major categories.

Frame Games

Frame games are inherently content-free, providing a structured series of interactions among players within which they may communicate information, experiences, and points of view. While some initial subject matter may be suggested as part of the introduction to the game, it is not critical to the conduct of the game. The interaction among players and the process it represents rather than the actual information is the real purpose of the game. Such games can be fairly easily "reloaded" with new information or even with an entirely new situation without changing the basic character or learning potential of the game. The reloading process itself may be an important research and learning experience.

Good examples of frame games are "They Shoot Marbles Don't They?" by Fred Goodman; "Sitte" by Gary Shirts; "Futures" by Olaf Helmer; "Impasse" by Richard Duke; and "Nexus" by Robert Armstrong and Margaret Hobson. A brief discussion of "Marbles" and "Sitte" will explain the nature of this game.

In "Marbles" players are asked to attempt a simple game of physical skill in which "shooting" marbles are used to try to hit "target" marbles subject to one or two rules about how and from where the shooter marble may be launched. Around this elementary set of conditions and associated payoffs, players are encouraged to build a society, including a government, a police force, a judiciary system, a public information system, and so on, reflecting to some degree their own circumstances. The nature of the society they create and the kind of laws that must be enacted to support that society become a microreview of the real world as the players themselves perceive it to operate. The game requires that players draw up a social contract, pass their own laws, elect or appoint leaders and peacekeepers, and ultimately stop and evaluate the kind of society they have created, judging its reality against the world they know and the world they aspire to. The experience is breathtaking and humbling, and the entire character of the society created comes from the players, not from the designer or the operator of the game. A careful analysis following the end of play yields important, fascinating, and sometimes painful insights about the individual's views of society and law and the role of government.

"Sitte" is a game that represents problems in coalition formation and communication in a typical multidimensional problem-solving situation. Players are divided into four teams representing different interest groups. A central decision-making body is identified (such as the common council of a city), and each team elects one person to represent their interests on this body. Players are confronted with seven issues reflecting upon their own team's interests differentially and five appropriation issues required to pay for the decisions made, each of which also affects the teams differentially. The players' only requirement is to agree on a decision and a fund-raising method, get both issues on the agenda of the decision-making body at the same time, and get them passed. If the decision-making body passes both the decision and the funding for it, the decision goes into effect and the status of the community and the teams changes accordingly. Each team has a limited amount of influence in each round in which it may vote for or against putting any specific issue or funding proposal on the agenda. Different interest groups must cooperate with each other to get issues and funding proposals on the agenda.

In most runs of "Sitte," few items ever reach the agenda, let alone get passed by the decision-making body. Within one hour or so, players come to understand the mechanics of the game and begin to ask themselves why they are unable to function within such a simple framework of decision making. The operator may ask how they expect to get anything accomplished in the real world when they are unable to function in the relative simplicity of the game. The problems clearly center around communication, trust, and cooperation. The problems are universal regardless of the details of the situation concerned. Reloading this game can become an instructive exercise in itself since it calls for identifying the major interest groups, the central decision-making body, and the major issues to be decided. If, after the initial run, the group attempts to reload the game to reflect its own situation, it has already come to grips with most of the major elements of whatever problem it met to address. This game can be an extremely powerful and instructive device in organizing and developing a sense of direction for a large, sometimes antagonistic and disoriented group of citizens meeting to work out a common problem.

Empathy Games

The major purpose of empathy games is to create an understanding of the position of some other person or point of view. Usually the players must take on the role of some other person or position and therefore begin to view first the game and then reality through the eyes of other persons. Application may be in areas of improving intercultural relations, in smoothing out interpersonal relations within an organization, or in helping public officials and social workers to understand better the needs and attitudes of their clients and publics. Good examples of empathy games include "Blacks and Whites" by Dov Toll; "Star Power" by Bud Crowe; "SimSoc" by Bill Gamson; "Baffa Baffa" by Gary Shirts; and "End of the Line" by Fred Goodman. Brief discussions of "Star Power," "End of the Line," and "Baffa Baffa" follow to illustrate this type of game.

In "Star Power," players are assigned to one of three groups and given a small number of colored poker chips. They are told that they may exchange poker chips with each other and that their status in the game will be determined by how many points they accumulate through holding various combinations of colors and numbers of poker chips. After an initial trading round, players begin to understand the relative value of the colors, and they also are made aware that certain teams have more points and better color combinations than others. The team with the most points is given veto power over rule changes in the game and assumes an increasingly dominant role in the play. Members of the other, less powerful teams tend to become more and more upset as the dominant team continues to accumulate wealth and power. The game ends when enough players are sufficiently angry to produce a spirited discussion of the meaning of the game and the meaning of status, wealth, and power. The game is provocative and can be used effectively in getting a group of people to address issues of this type. Some caution is called for in handling interpersonal feelings.

"End of the Line" was designed to teach workers in social service agencies about the feelings and problems that arise as people get older and more dependent on others for daily assistance. Through a series of clever physical and intellectual requirements, players experience a loss of mobility, eyesight, and memory, being forced gradually into positions of helplessness and total dependency as the game progresses. Other players, representing social service agencies, attempt to provide assistance to those most in need, but they are continously hampered by lack of funds and inadequate information about exactly who needs help and what kinds of assistance they require. Periodically during the game players "die," and if they have not managed to play a minimally successful strategy during the previous rounds, they either die or lose more of their faculties. Near the end of the game, many of the "old people" in the game are relegated to sitting idly in their chairs waiting for periodic handouts from the social service agencies in order to stay alive. Some choose to die instead. The game can be depressing, and users are advised against having old people play. Persons who have played claim to be much more aware of the needs and attitudes of old people and to be much better prepared to work with them.

"Baffa Baffa" was commissioned by the U.S. Navy to make naval personnel more sensitive to the cultural beliefs and customs of the various countries they might visit. In the game, players are divided into two subgames, each of which is played with substantially the same rules, subject to a few minor variations in each. In both situations, players trade poker chips as in "Star Power." In one game, however, points are awarded according to how the transaction is carried out, how many compliments are exchanged, how much good feeling is expressed, and so on. Also in the game, males may not trade with females without the permission of the eldest male in the game. In the other game, points are awarded to the best combination of colors as in "Star Power," and players are not allowed to speak or touch each other while trading. Once players in each game have begun to understand how their own game operates, a few visitors are allowed to move from one game to the other and to try trading in the other game. Serious difficulties arise almost immediately in both games; the visitors are viewed as boorish, hostile, dumb, or insensitive, depending on the culture they are from and the one they are visiting. After several such visits back and forth, the two groups are brought together and asked to give their impressions of the visitors from the other culture. A fruitful discussion of value differences between seemingly similar cultures opens the door to a better understanding of how to behave more sensitively in other societies.

Resource Allocation Games

Players begin by competing for shares of one or more resources, which are in relatively scarce supply, such as land, money, water, food, or power. Although the games inevitably have a competitive flair, most of them result in players' realizing that some form of cooperation and planning will usually produce more usable resources for all. In general, such games introduce planning and cooperation concepts indirectly through creating a demand for such activities among the players themselves. Good examples of such games are "CLUG" by Allan Feldt; "City IV" and "River Basin" by Peter House; "SNUS" by Rob Carey and Richard Duke; "New Town" by Barry Lawson; "Urban Dynamics" by Larry McClellan; and "WALRUS" by Allan Feldt and David Moses. Brief discussions of "CLUG" and "Urban Dynamics" follow.

"CLUG" (Community Land Use Game) makes each player or team of players a private real estate developer and operator. They may buy land through competitive bidding, develop it into one of five basic land uses, and operate it at whatever profit or loss the sum of their business decisions yields within the structure of the game and the actions of other players. Major elements of an economic system are represented in the form of costs for land, labor, and capital plus costs incurred for taxes and transportation of goods and labor to the marketplace. Shortage of land at prime locations soon forces players into reconsidering their positions and undertaking efforts to control and plan present and future development more rationally. The game has no clear-cut winners. Most players find that they are actually competing against the system and the environment represented by the game rather than against each other.

"Urban Dynamics" reflects an urban setting similar to the regional setting portrayed by "CLUG" except that political and social forces play a much more explicit role in the outcome of the game. Individual teams of players represent major ethnic groups and power blocks in Chicago in the 1920-1950 period, including an increasingly large and militant black population, which to a substantial degree is cut off from sharing in the jobs and opportunities of the city. Once the basic economic components of the game are understood, much of the play revolves around demands for housing and political power by the various interest groups represented. As the teams representing lower-income and black groups become larger, the entrenched white business leaders are forced to find ways of meeting their demands for jobs, housing, votes, and so on. The game is particularly useful in portraying the interplay of social, economic, and political forces in shaping the historical development of a city.

"SNUS" (Simulated Nutrition Simulation) was developed for regional and national officials concerned with the balance between economic development and the nutritional well-being of populations in developing countries. Players represent officials of regions within a developing country and must make choices about how to use the limited resources available to them: growing more food, engaging in more internal or international trade, building housing or factories, and so on. Each round is followed by a report on the state of the regional and national economies and an opportunity to change national and regional policies in an effort to improve the social and physical level of the population and the economic position of the individual regions and the country as a whole. As in the real world, a choice must often be made between steel mills and bread.

Process Games

In these games players learn a certain number of critical steps that must be taken to play the game successfully. These steps and the way they interrelate with similar steps being taken by other players represent some important form of political, managerial, legal, or other form of process that the game is designed to represent and teach. Examples of such games are "Metropolis" by Richard Duke; "Metro-Apex" by Richard Duke, Dick McGinty, and others; "GSPIA" by Frank Hendricks and others; "WARD" by Larry Coppard; "The HEX Game" by Richard Duke; "The State Legislative Game" by James Coleman; and "Woodbury" by Marshall Whitehead.

In "WARD" players represent local neighborhood residents interested in developing a neighborhood revitalization program with federal funding such as was available during the late 1960s and early 1970s. Different groups of players representing different neighborhoods are given fairly detailed descriptions of their neighborhoods, including building condition, social characteristics, demographics, economic factors, and so on. They are asked to develop and seek federal funds for a neighborhood redevelopment program and are required to carry out a series of steps and activities to get their program approved. These steps are a simplified but exact representation of the procedures required by federal regulations in the late 1960s. The players learn what they would have to do in reality to get a program for their own neighborhood approved, a problem of considerable complexity at the time due to the byzantine structure of federal regulations and procedures built into the neighborhood redevelopment program. The game is effective in training local citizens in how to develop and gain approval for their own development program, although federal funding of this program was dropped about a year after the game was developed.

In "Metropolis," players are divided into teams of politicians, developers, and planners in a metropolitan region. Politicians, each representing one of three different wards, are required to set a tax rate for each round. With a portion of the revenues collected, they must decide upon the level of capital expenditures the city will make each round in each ward in response to demands for new services and facilities announced in a newspaper printed for each round. Planners are required to recommend new capital projects to the politicians two rounds in the future and to estimate the amount of tax revenues that will be available for such projects. Developers must invest their capital in new developments in each ward in anticipation of where and what type of new capital investments will be made by the city. Politicians win or lose reelection according to how well they meet the demands of their respective wards. Planners retain or lose their jobs according to how many of their recommendations are accepted. Developers win or lose money according to how well they anticipate or influence the capital developments made in each ward. Play continues for five or six rounds until all players come to understand the capital budgeting and planning process and the political and economic arena within which it operates.

In "HEX," players represent local regional and national officials in a developing nation. Local officials try to balance and develop their local economies, including trading with their immediate neighbors. Regional officials try to develop a regional economy, including interregional trade and the development of specialized industries and goods based upon the unique properties of local areas. Meanwhile the national officials are primarily concerned with international trade and finding a position for their own nation within the international economy, seeking the development of local goods to provide a good balance of trade for the nation with other nations of the world. Each of the three levels is faced with its own problems of balancing the economic and social development of its population, and the need for communication and cooperation between levels is obvious. During the game players begin to realize the kinds of problems and conflicts that arise within and between levels that inhibit good communication and cooperation and, thereby slow or prohibit economic development of the society. Officials in many developing countries that have used the game have found it to be useful in developing better communications between officals at varying levels of government.

DECISION MAKING

Planners for the most part assist in decision making rather than make the actual decision. The mayor, council, city manager, and others who make decisions base these on a variety of sources, of which the planner can be one of the most important. One might expect that by the nature of decision making, some analytical methods exist that planners can use to make a case for either implementing or not implementing a certain decision. There is, in fact, a large body of literature on the nature of decision making using some probabilistic techniques for selecting various choices. The techniques that use probabilistically determined outcomes are called *Bayesian*. Three of these techniques stand out as the more common ones for planners: decision trees, contingency tables, and minimax (or maximin) analyses. These techniques were developed under the relatively new engineering discipline of operations research to assist in making rational choices based upon probabilistic associations for multiple outcomes.

The decision tree is probably the most familiar technique. It uses the conditional probabilities associated with events that may be linked to a number of outcomes. The final probability of the linked outcomes can then be determined by multiplying the various outcomes along the branches to determine a final probability for a string of events. For example, suppose the mayor of Middlesville wants to know what effect ignoring the garden club's complaints will have on the upcoming city election. At this point, he knows that his opponent has as good a chance of winning as he does and that it is also a fifty-fifty possibility as to whether he will do anything about the downtown trash problem before the election because many problems need his staff's attention now. A quick analysis shows that if he does not do anything about the trash problem, 90 percent of the garden club will oppose his reelection, which can greatly affect his chances. In fact, if they resist his election, his odds of winning will drop from 50 to 40 percent. The question he needs to answer is whether to respond to the garden club now or wait until after the election.

A decision tree for this type of problem is illustrated in Figure 4.6. In this simple example, Figure 4.6 shows that the mayor's chances for reelection are decreased if he does not respond to the group, and it

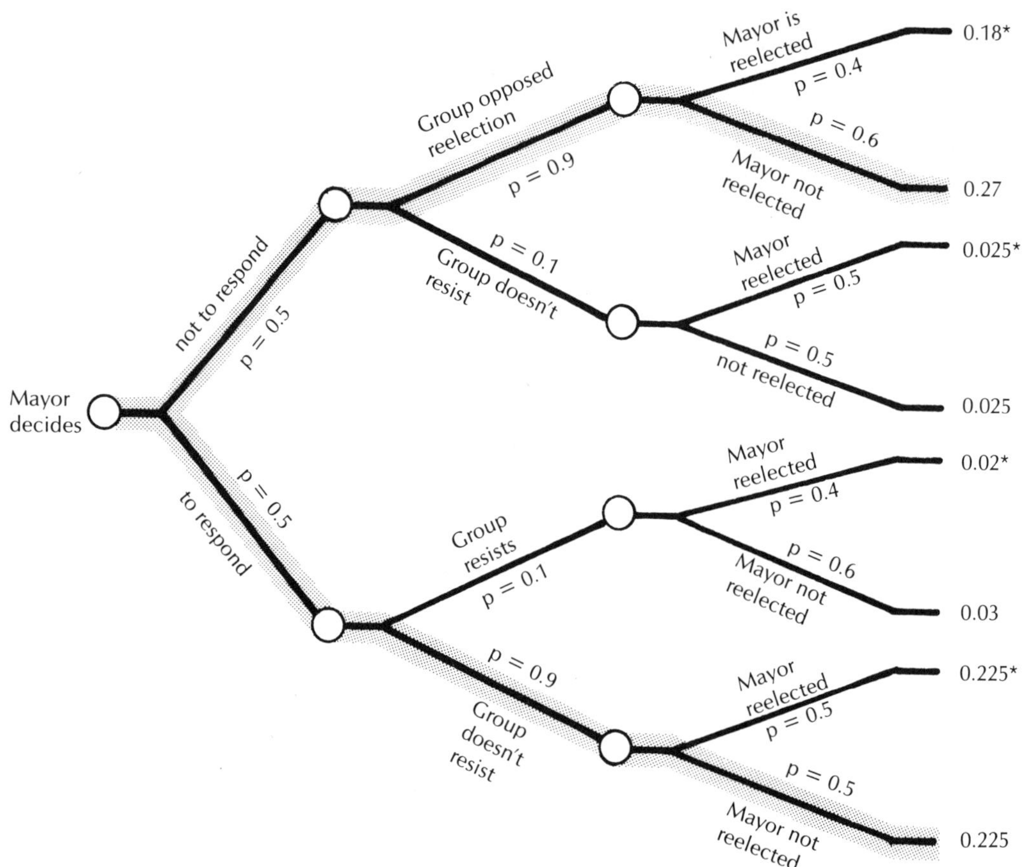

Figure 4.6 Decision tree of Mayor Lorch's probability of reelection.

resists his election. This results in a 0.27 probability of the mayor's not being elected and compares to a probability of 0.225 of being elected even if he does respond to their request. Indeed he has found that as a result of the garbage issue, his chances for reelection have changed from 50 percent to only 45 percent, which is calculated by totaling the reelection probabilities on the decision tree. Therefore it is in the mayor's best interest to maximize his chances of winning by responding to their request immediately.

The example contains a number of statistical assumptions that must be validated before any action takes place; however, it demonstrates the use of probabilities in assisting decision makers. The necessity of establishing the independence and mutual exclusiveness of the events is beyond the scope of this book, but such methods as contingency tables can, in part, be used for this purpose.

Other decision techniques, such as minimax and maximin strategies, are designed to allow the decision maker to select strategies that will, respectively, maximize the chance of getting at least the minimum benefit or minimize the chance of expending the maximum costs associated with program selection. Numerical data are analyzed either by inspection if the data are simple enough or by linear programming techniques that calculate the optimal approach based upon the initial conditions.

The other techniques described in this chapter are all used in the decision-making process since they communicate information to the decision maker in ways that quantitatively or qualitatively lead to some rational decision-making criteria. The methods detailed here, however, are exemplary of the most commonly referenced Bayesian decision techniques. Such methods, where the probability of independent events is determined leading to the calculation of a probability of the outcome of a string of events, would be ideal if all events were indeed independent and the probability of occurrence for each event were well known. But this is not the case, and frequently probabilities have to be assigned based upon some heuristic method, such as the opinions of experts (called Delphi techniques) or upon best guesses by an experienced observer.

Various methods for determining the probable outcome of a future event are available to those who assist in decision making for them to add credence to the acceptance or rejection of a decision. The inferential statistics discussed earlier in this chapter are available as valid methods to assist in making complex decisions; however, making decisions under uncertainty is probably the most common problem facing the decision maker concerned with the future. (See Lindblom, 1959, for a less rigorous account of decision analysis.) As a result, most of the mathematically rigorous methods are not as useful as one would hope. At least four nonmathematical factors act as the determinants of an outcome for any decision: risk, turbulence, uncertainty, and change.

Risk

Risk analysis has been developed in an analytical fashion, and some risk assessments can be performed on certain types of events. These types of assessments usually relate to technical events and involve the probabilistic assessment of a string of independent events. Analyses such as the risk of a meltdown at a nuclear power plant can be made in this fashion. The risk may change, however, as new factors are introduced or other analyses taking different factors into account are performed, so it is not always easy to come up with one probability associated with any single risky event. But more importantly, the decision maker must decide how much of a risk he or she is willing to take in making a decision. For example, if the probability of a serious nuclear accident at a power plant is 1 in 20,000 over a one-year operating period, then it may appear that it would be a safe decision to approve the construction of a nuclear power plant. However, if one further considers that there might be 200 nuclear power plants operating over a twenty-five-year period, then the probability of a serious accident is 1 in 4 in that period. But if a serious accident is defined like the accident that occurred at Three Mile Island in 1979, then the risk may be worthwhile if the benefits of the additional nuclear energy outweigh these risks. On the other hand, the United States currently has more electric generating capacity than it needs, so no new plants would be of benefit. But the majority of the operating plants burn coal, oil, or natural gas, fuels that either harm the environment when they are burned or are extracted, or are in dwindling supply.

This argument can go back and forth without the benefit of a rigorous numerical analysis even though quantifiable information exists. The risk that a decision-making body will ultimately take depends in large part on previously taken risky decisions and the consequences of those decisions. Those assisting in decision making can present all the arguments and even numerically assess some of the risks associated with the decision, but the actual decision for the most part will be based on the decision makers' personal experiences.

The planner should learn the risks decision makers have taken in the past and either state them or be able to address them when presenting data. If the city of Middlesville's Environmental Engineering Department previously refused to pick up garbage downtown on weekends because of the fear of accidents during heavy weekend traffic, then Junior Planner should be prepared to address this issue when he makes his recommendation for a weekend pickup, no matter how rational his other reasons for making the recommendation may be.

Turbulence

A turbulent environment can be caused by a single event such as the evacuation of a community due to a natural or human-caused disaster or a combination of events—for instance, the transition of a community from one type of industrial base to another. A high level of turbulence is usually perceived as a negative factor, so decisions that can increase turbulence, no matter how short-lived that turbulence might be, will be avoided in general. Consider the clearing of an area having a particular land use such as residential to accommodate another land use, such as commercial or industrial. This type of decision will create turbulence for the residents and will also create resistance to making the final decision. Turbulence is not necessarily bad, but it can create problems if the decision in all other respects is an appropriate one for the entire community. Those who have avoided making decisions because they might create turbulence, no matter how short term, in all likelihood will continue to avoid such decisions.

Planners must be aware of turbulence-generating decisions and present their information by showing that short-term turbulence may be necessary for the overall benefit of the community, and a much more harmful, long-term turbulence may result from not making a decision. The decision to remove the residents from Love Canal in New York to avoid any further biological harm caused by exposure to toxic wastes created a turbulent situation, but the alternative of having the families stay in the area was potentially far more turbulent.

Uncertainty

Even with good probabilistic analyses, a certain amount of uncertainty regarding any decision is always present. Each of us deals with uncertainty in our own way, and some of us are willing to accept far greater levels of uncertainty than others. Furthermore, the type of outcome associated with any decision will also determine how much uncertainty one might accept. If a particular outcome, such as a nuclear accident, no matter how small or localized, is totally unacceptable to a decision maker, then any amount of uncertainty associated with the possibility of a nuclear accident will not be accepted. On the other hand, if one is willing to accept a highly improbable, low-level nuclear accident, that person will also accept a certain level of uncertainty associated with a nuclear plant site.

Planners who are assisting decision makers will also have to understand their own personal level of acceptance of uncertainty, but more important, they must be fully aware that a decision maker or a decision-making body will have its own unique level of acceptance. As a result many planners are frequently frustrated by their inability to convince a decision-making group to accept or reject decisions based on what the planner feels to be an acceptable level of uncertainty.

Change

Almost all policy decisions lead to some change in either the behavior or appearance of the affected group and its environment. A fundamental premise of complex systems analysis is the system's ability to resist change no matter how valid or appropriate that change might be. Like uncertainty, the amount of change one is willing to accept varies from each individual and from each group. The difference is that change is more qualitative in nature, and the perceived change that may result from any particular decision is highly individualistic and frequently value laden. In many instances, prejudice and bigotry on the part of the decision maker will have a direct bearing on the type of change he or she perceives and is willing to accept. Zoning changes that appear to give some continuity to a community may be proposed to prevent perceived or real ethnic changes in a neighborhood. In fact, one of the first zoning ordinances passed by a city was to exclude laundries from San Francisco's Nob Hill area. The reason was not for environmental purposes, as some argued, but to prevent Chinese people from moving into the area.

Planners must be aware of the broad spectrum of changes, real and imaginary, that may result as the outcome of any decision and be prepared to address them. This may prove uncomfortable for and at times antagonistic to some decision makers, but unless the real and perceived changes are expressed, rational arguments may not lead to a desired decision.

PLANNING MANAGEMENT

Planning analysis is not just numbers and equations and methods for deriving them; frequently it requires synthesizing information solely on the basis of how the information is presented. Indeed the initial analyses of most problems are based on information presented in a format that best suits further analysis. Management problems involve complex interactions among people, tasks, and productivity, and usually they can be presented or arranged in such a way that a logical flow through can be analyzed.

Various planning management methods are described in the literature; they include PERT (program evaluation and revision techniques) and CPM (critical path method). In these techniques, a collection of tasks is specified that, when complete, leads to a final outcome, which can be a program, a manufactured product, or a complex system. The project flows from task to task, with completion time estimated for each individual. Tasks are also represented leading to an overall assessment of how critical each task is and how each fits into the final outcome. These methods are extensively used in planning as are the techniques of planning, programming, and budgeting system (PPBS), matrix representations, and similar methods whereby information is synthesized into a display useful for large program management. The bibliography at the end of this chapter covers these and other techniques in detail. Two examples of how such methods may be employed to communicate planning problems are presented here.

Most complex management plans require identification of non-quantitative processes, such as the various tasks to be accomplished, and who might do them, along with some fixed quantifiable variables, such as the time or money it takes to complete the tasks. Suppose that Junior Planner from Middlesville had to submit a program management plan to his planning director prior to initiating any other actions on the downtown trash problem. He might present the chart shown in Figure 4.7 indicating the time and effort expected to complete each task. Charts such as these, called *milestone charts*, are common management tools used to establish the overall cost and effort needed for completion of budgeted programs. They serve to check on the progress of the program and are useful in indicating potential cost and time overruns.

Another method for presenting information that guides managers in implementing an overall plan is in the use of a two-dimensional matrix presentation where some critical information about management needs and ways to meet them is given in each cell. We recently completed an energy planning and management program using this type of matrix for the U.S. Department of Interior, which will be used for parks and recreational areas. (See Figure 4.8.) The energy needs of various facilities, such as outdoor lighting, vehicle fleets, and swimming pools, are listed vertically on the matrix, and the fuels used to meet those needs are listed horizontally. The primary purpose of the program was to manage the park's energy budget so as to minimize the energy used while maintaining a high level of services. The information contained in each cell of the matrix summarized the cost and savings attributable to a particular conservation strategy applicable to the fuel used at the facility listed.

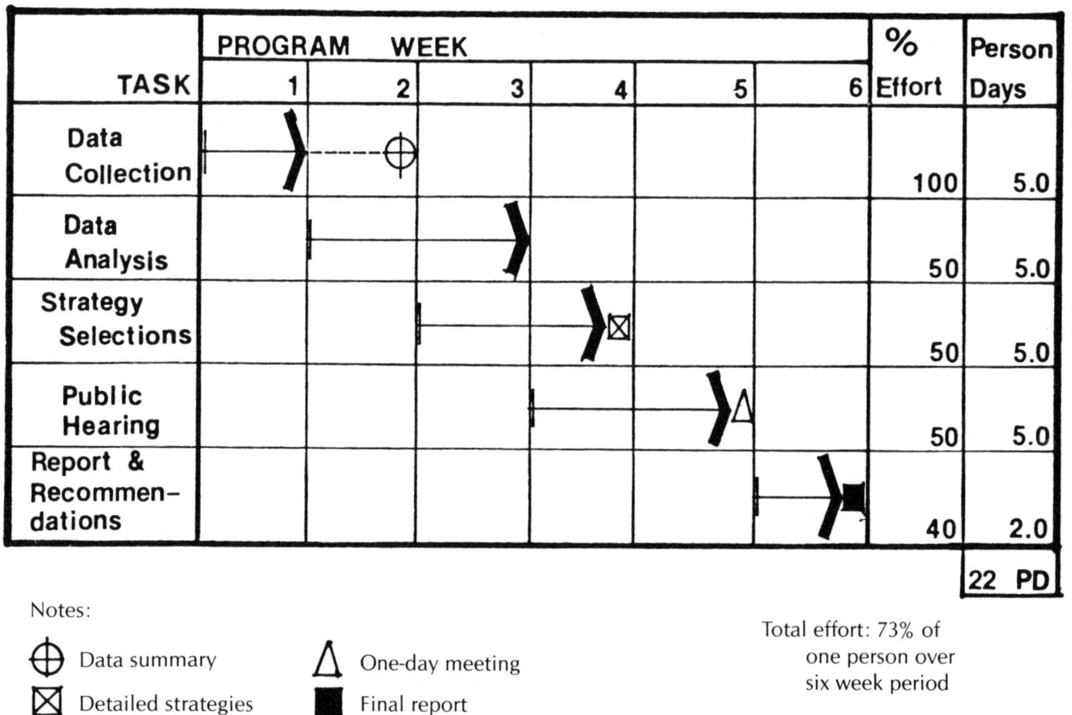

DOWNTOWN TRASH ANALYSIS
PROGRAM WEEK
TASK
1
2
3
4
5
6
% Effort
Person Days
Data Collection
100
5.0
Data Analysis
50
5.0
Strategy Selections
50
5.0
Public Hearing
50
5.0
Report & Recommen-dations
40
2.0
22 PD
Notes:
Data summary
Detailed strategies
One-day meeting
Final report
Total effort: 73% of one person over six week period

Figure 4.7 Program milestone chart.

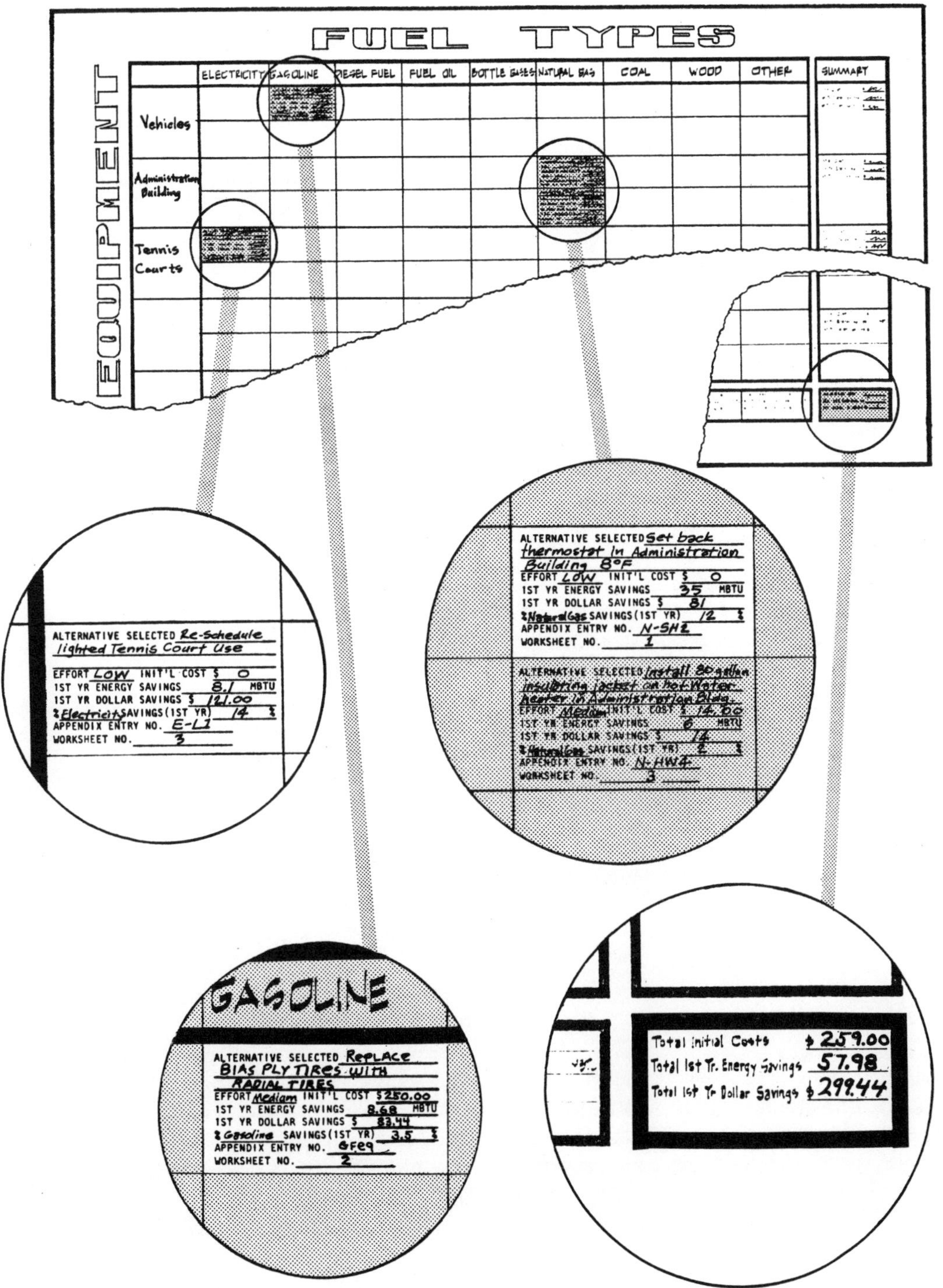

Figure 4.8 Energy planning matrix.

When completed, the matrix contained a large variety of strategies for use in developing an overall park energy management plan. The information for each facility was summarized at the end of each row, and the information for each fuel source considered was summarized at the bottom of each column. The summary information established an overall cost and the potential energy savings for the total implementation of all the strategies. The method is particularly valuable because it allows for the presentation of qualitative information, such as the political and institutional characteristics associated with implementing each strategy, by a color-coding process along with the detailed quantitative information. In this manner the park manager could select which strategy should be implemented and in which order implementation should occur. Since the costs and benefits of implementing strategies are cumulative, strategies having high costs and marginal benefits could be implemented at a later stage of the program, or even not at all if the overall goals of the program are reached with the higher priority strategies.

Planners who use such management tools for implementing programs should be aware that there is a lot of room for individual development of other methods. No one method is ideally suited for all management programs, and the creativity of the planner who is drawing upon methods developed by others is always challenged in developing new management programming tools.

FORECASTING

When planners discuss future events, they usually do this in terms of alternative futures. No unique future exists now, of course; rather a collection of probable futures exists, and by proper planning and appropriate decision making, a group can prepare for the futures with higher probability of occurrence.

The history of modern forecasting used by planners is relatively new, dating from around the end of World War II. Three primary reasons are usually stated as to why forecasting is used in planning: problem identification, consequence of actions, and normative judgment.

By observing trends and patterns over time, problems will surface that may not have been obvious without some type of forecasting process. By the same token, one can assess the consequences of one's actions not only by evaluating the primary effects but also by the secondary and higher-order effects of a particular action, which can be discerned over time. Various forecasting methods are quite responsive to these higher-order effects, and without them such effects would be almost impossible to predict. Finally, the outcome of decisions should lead to goals that are normative—in other words, within the accepted norms of the group affected by the decisions. Hence by appropriate forecasting, problems can be identified, the consequences of certain decisions surrounding those problems evaluated, and those decisions with the highest probability of leading to acceptable goals can be implemented.

Forecasting methods range from examining a single variable over time to complex multivariate interactions. The simplest types of forecasting usually involve only one variable, and based upon the patterns or trends observed, one can forecast the future behavior of that variable. More complex forecasting techniques generally include a number of variables, and either through some process or by using computer techniques designed to examine a large number of interactions, future events can be simulated and assessments made. These methods usually require quantifying the variables being examined—number of houses available in ten years, vehicle density in five years, number of different types of energy sources available in twenty years, amount of air and water pollution as a result of industrial growth in the next five, ten, and fifteen years, and so forth. But forecasting can be qualitative as well. Various methods are available that can offer some understanding of how the values of a group may alter future responses to various public and private acts. The current public awareness of environmental degradation caused by noncontrolled industrial processes is one example of how a change in societal values toward industrial growth can affect product cost and product availability of a region.

The *Handbook of Forecasting Techniques* (Mitchell et al., 1977) details a large number of available forecasting techniques. An examination of a couple of the techniques will demonstrate how they might be used. Scenario generation is a common forecasting technique that is used to present different pictures of the future as a result of varying outcomes of major events. Scenarios can be both qualitative and quantitative. They can be generated from either the planner's knowledge of the problem or by random assignments to the descriptive information. An energy scenario for the year 2000 (Figure 4.9), along with a collection of other scenarios, each having different values assigned to the descriptive variables, can then be used in combination with other techniques to arrive at some U.S. energy forecast.

A method often employed with a collection of scenarios is the Delphi process. In this technique a panel of experts examines a range of scenarios similar to the one shown in Figure 4.9 and assigns some probability to the outcome of each scenario. The resultant probabilities are then shared by the panel who generally do not know which members assigned the probabilities to the various scenarios. After seeing all the probabilities for each scenario, the expert is asked to reexamine his or her original assignments and either to substantiate why they are different from those of the other experts or to change them, if he or she so chooses. Finally, a consensus probability is calculated from all the experts' contributions for each scenario, resulting in a probabilistic forecast for the entire set of alternative futures.

Complex systems analysis using computers to handle vast amounts of data or calculations is also used to forecast future events. Forrester's *Urban Dynamics*

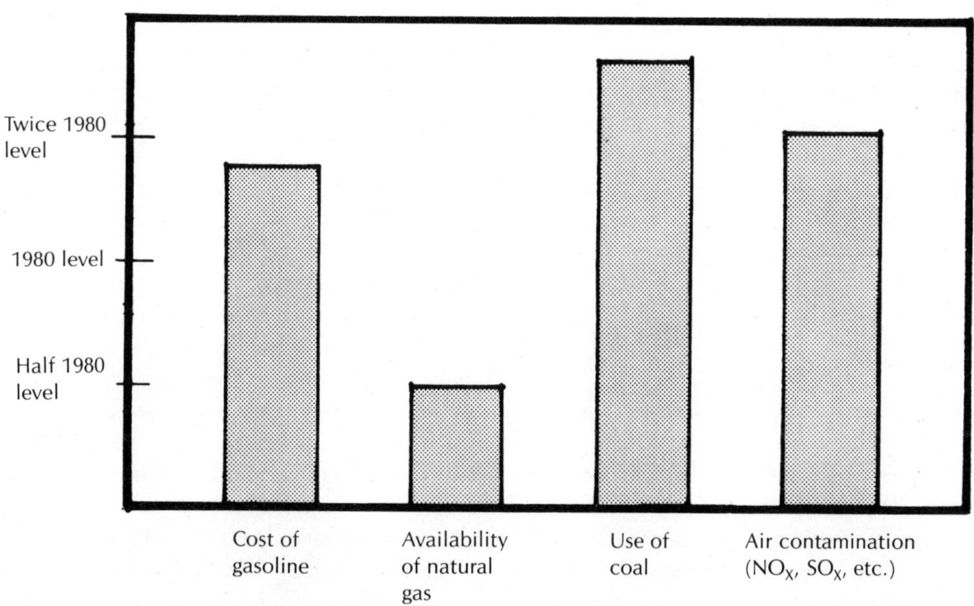

Figure 4.9 U.S. energy scenario, year 2000. In the year 2000 the cost of gasoline has almost doubled over 1980 costs, and the availability of natural gas has decreased substantially. As a result the use of coal has dramatically increased without benefit of extensive pollution controls, leading to a considerable increase in environmental deterioration.

(1969) is probably the most ambitious example of this technique. He took into account factors associated with housing, jobs, and industry to develop a complex model of urban interactions. By varying the rates of control variables, such as the rate of new job generation, changes in worker housing, and worker migration rates, forecasts for the future needs of cities in terms of new housing, new jobs, and industries could be generated.

Simulation/gaming also lends itself to forecasting. By having the players enact roles and make decisions in their roles, future events can be simulated. The knowledge of how decisions affect future outcomes in a game situation is easily extended to a forecast, if the underlying restrictions of the game model are kept in perspective.

Whichever forecasting technique the planner uses, a number of cautions must be observed. The most obvious is that the further into the future one forecasts, the more improbable the forecast. Long-range forecasting is not the most accurate tool for policy making, and planners should be aware of the accuracy of each technique. Another problem with certain types of forecasting is erroneously assuming causal relations between events simply because a correlation exists. Finally a forecast is only as good as the data used. Limited amounts of data, poorly measured data, and hastily reduced data will lead to inaccurate forecasts no matter how sophisticated the forecasting technique. Computer people refer to this as "GIGO" (garbage in; garbage out).

APPLICATIONS

1. From the census tract volume for a nearby city of about 100,000 population, compute the percentage distribution of total population and black population among census tracts. Do the two populations appear to be evenly or unevenly distributed?
2. Draw a map illustrating the distribution of black population among the census tracts. Compute the coefficient of dissimilarity between the total and black populations, and discuss its significance. If available, compute the same figures for the 1970 population and consider the significance of changes in the distributions over the past decade.
3. You have been assigned as the program manager to evaluate various alternative energy sources for the state of Hiatonka. The evaluation is to be presented to a state commission in nine months. Develop a milestone chart for your program, listing each task and the estimated times and costs to complete them. Although some of these tasks may be highly technical or analytic in nature, assume that you or your staff are capable of completing them.
4. As a planner, you have been asked to assess the impact of locating an adult bookstore near a middle-income, residential neighborhood. A zoning change has been requested, and you are to prepare the city's position for or against the change. Examine the elements of risk, uncertainty, turbulence, and change that might be expected as a result of the proposed zoning change.
5. Run a simple game requiring fifteen to thirty minutes on a group of ten to twenty people. The game is called People Weaving.

   Place a series of parallel marks about twelve inches apart on the floor with six-inch pieces of masking tape. Place one more strip than there are people to play the game. Leaving the strip in the middle empty, have the players each stand on one strip facing toward the empty center strip. Give them the following rules:

   a. They may not talk but must communicate through gestures.
   b. They may move only forward, not backward.
   c. They may move only onto an empty strip.
   d. They may step around another person only if that person is facing them before the move.
   e. The object of the game is to exchange posi-

tions fully from each side of the center, so that each group is then standing with their backs toward the center empty strip.

f. When further "legal" movement is impossible, they may all raise their hands over their heads, return to their original positions, and try again.

After the group has successfully woven itself through the center position into the winning position, have them turn around facing each other and ask them to reweave themselves back into their original positions. Note and comment upon the changed position and responsibilities of those now in the center who used to be on the ends. Most groups find the only solution to this problem in fifteen to twenty minutes after a dozen or so tries. People in the center tend to be very active while those at the ends tend to become bored and restless.

Ask the players to describe their feelings about the ability of their group to solve the problem. Who did the work? Why did they and not others do it? How did they feel when someone took a wrong step and jammed the pattern so they had to start over again? Why were people at the ends not more deeply involved in the problem at first? How did they feel when people in the center tried to tell them what to do? How did they feel when suddenly they were in the middle and had to restart the weaving? Are there groups in the community in a similar position? Would the game have gone faster if they could have talked to each other? What would they have said and who would have said it? Who won the game?

REFERENCES

Forrester, Jay W., 1969, *Urban Dynamics,* MIT Press, Cambridge, Mass.

Lindblom, Charles E., 1959, "The Science of Muddling Through," *Public Administration Review* **19**:79-88 (Spring).

Mitchell, Arnold, Burnaham H. Dodge, Pamela G. Kruzic, David C. Miller, Peter Schwartz, and Benjamin E. Suth, 1977, *Handbook of Forecasting Techniques,* 2 vols, INR Report 75-77, National Technical Information Service, Springfield, Va.

BIBLIOGRAPHY

Barclay, George, *Techniques of Population Analysis,* John Wiley, New York, 1958.

A good summary of basic demographic measures and techniques. No direct applications to urban planning are given.

Bogue, Donald J., "A Composite Method for Estimating Postcensal Populations of Small Areas by Age, Sex, and Color," *Vital Statistics—Special Reports,* **47**(6):167-185 (August 1959).

An early and clear discussion of this very useful technique for estimating intercensal populations.

Browing, Harley, "Methods for Describing the Age-Sex Structure of Cities," in Jack Gibbs, ed., *Urban Research Methods,* Van Nostrand, Princeton, N.J., 1961.

A good discussion of the construction and interpretation of population pyramids.

Coppard, Larry, and Frederick Goodman, eds., *Urban Gaming/Simulation,* School of Education, University of Michigan, Ann Arbor, Mich., 1979.

An overview of gaming/simulation together with descriptions and discussions of about one hundred games dealing with urban problems.

Dickey, John W., and Thomas M. Watts, *Analytic Techniques in Urban and Regional Planning,* McGraw-Hill, New York, 1978.

An excellent introduction and overview of the various analytic methods used by planners. Provides useful appendixes on basic mathematics and computer languages.

Duncan, Otis Dudley, and Beverly Duncan, "A Methodological Analysis of Segregation Indexes," *American Sociological Review,* **20**(2):210-217 (April, 1955).

Thorough and clear coverage of the coefficient of dissimilarity, Lorenz curves, and the Gini concentration ratio.

Forrester, Jay W., *Urban Dynamics,* MIT Press, Cambridge, Mass., 1969.

A detailed systems analysis of an urban area with projections of change in employment, housing, and industry based upon a complex but limited computer simulation model.

Isard, Walter, *Methods of Regional Analysis: An Introduction to Regional Science*, MIT Press, Cambridge, Mass., 1960.

Provides a major review of location quotients, economic base calculations, and related economic indicators.

Krueckeberg, Donald A., and Arthur L. Silvers, *Urban Planning Analysis: Methods and Models*, John Wiley, New York, 1974.

A good, mathematically rigorous text and reference work. Examples appropriate to planning are used throughout.

Lindblom, Charles E., "The Science of Muddling Through," *Public Administration Review* **19**:79-88 (Spring 1959).

The first statement of the now well-known and widely accepted critique of scientific decision making.

Mitchell, Arnold, Burnaham H. Dodge, Pamela E. Kruzic, David C. Miller, Peter Schwartz, and Benjamin E. Suth, *Handbook of Forecasting Techniques*, 2 vols., INR Report 75-77, National Technical Information Service, Springfield, Va., 1977.

A detailed description of thirty-one forecasting techniques. A practical and well-formulated handbook, with examples applicable to environmental, social, and economic forecasts.

Raiffa, Howard, *Decision Analysis: Introductory Lectures on Choices Under Uncertainty*, Addison-Wesley, Reading, Mass., 1968.

A sound mathematics text on decision theory. Readable for nonmathematicians with some statistics background and recommended for planners interested in decision-making analysis.

Shryock, Henry, and Jacob Siegel, *The Methods and Materials of Demography*, U.S. Government Printing Office, Washington, D.C., 1975.

A thorough and detailed coverage of several methods of population estimation and projection, plus most other standard demographic techniques.

Stadsklev, Ron, *Handbook of Simulation Gaming in Social Education*, 3 vols., Institute of Higher Education Research and Service, University of Alabama, University, Ala., 1980.

A detailed and complete coverage of most available games together with a textbook on gaming aimed primarily at secondary school teachers. Volume 2 covers manual games, and volume 3 covers computerized games.

5

Working with Small Groups

Peter Ash

The members of the private consulting firm, having just signed a contract with the states of Hiatonka and East Victoria to survey energy options, treated themselves to a self-congratulatory lunch. Over coffee, they discussed how to proceed. The Pleasant River constituted a 200-mile border between the two states, and the river could perhaps be used for hydroelectric power. The senior members of the firm thought that a water group should be formed to explore this possibility and that A. Planner should be the leader. They decided that some governmental representatives should be in the group, as well as several planners. The consultants were breaking their task down into components and did not want to specify the group any further; they were confident the leader, in whom they had great faith, could take care of the details.

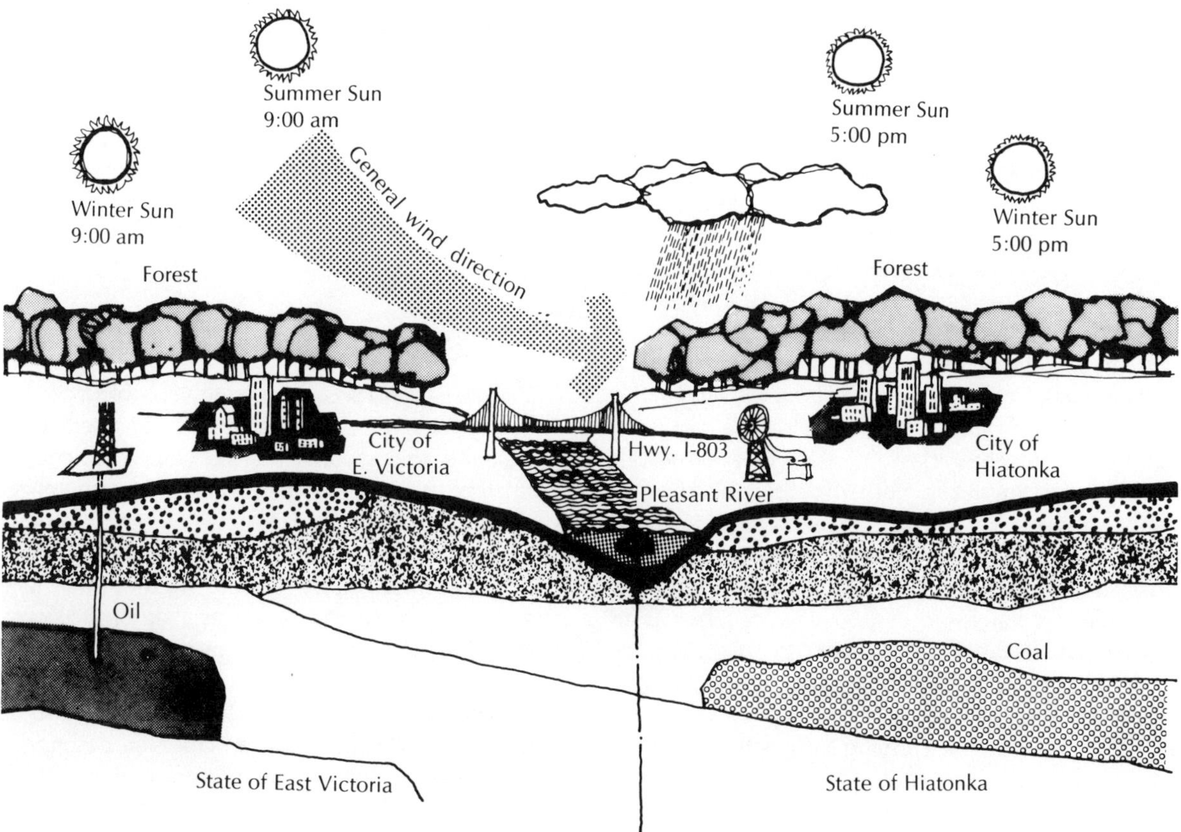

Energy resources in Hiatonka and East Victoria.

Groups like the Water Group are common in planning. This chapter considers issues relevant to the effective use of small groups by planners. It will focus on small groups of fewer than ten members, meeting repeatedly with a fairly stable membership, with a leader clearly defined, and with a task articulated at the outset. The emphasis will be on how to work with such groups rather than on the abstract theory of groups. The Water Group will provide illustrations of many of the techniques.

GROUP PROCESSES AND THE GROUP TASK

Tasks

Planning groups are work groups; they are set up to fulfill particular tasks. The task of the Water Group is to produce a report that surveys options for damming the Pleasant River and evaluates the consequences of these options. The report will go to the planning consultant firm and to the heads of the departments of energy in the two states of Hiatonka and East Victoria. The consultant planning firm at lunch had a different task—to decide what steps were necessary to survey energy options—and one of their decisions was to create the Water Group. A group may have a highly technical task—to do a technical survey of a dam site, for example. Planners also use groups to carry out nontechnical tasks, such as achieving greater cooperation among members of the planning firm. In such groups, the outcomes are not decisions but rather attitudinal changes among planners that will facilitate their working together. Another type of attitudinal change group is one whose task is to develop acceptance for an already determined plan. There are many public hearings that have this task as a covert function.

Whatever the task, carrying it out is the job of the group. To the extent that the task is not clear to the group or to the extent a group does not accept its task, the group cannot do its work. The group task can be broken down into smaller problems. The following schema is useful for this purpose:

1. Stating the problem
   a. Exposition of the problem
   b. Delineating objectives of a solution
2. Clarifying issues
   a. Clarifying options, for both information gathering and final actions
   b. Clarifying external events that may occur
   c. Judging consequences of various courses of action
   d. Estimating probabilities of uncertain events
   e. Clarifying preferences
3. Choosing among alternatives
   a. Comparing outcomes of options
   b. Making trade-offs among values
   c. Reporting the decision

These steps may be iterative. For example, clarifying preferences and choosing among alternatives may lead to changing the definition of objectives in the original problem statement and raise new issues that need to be clarified.

Process

When a group is carrying out the problem-solving steps, members are interacting. These interactions constitute the group process. Group process is sometimes discussed in terms of an individual member's actions, motivations, and feelings toward both other members and the group as a whole. Group process is also conceptualized by considering the group as a whole. Groups seem to take on a life of their own, which becomes more than the accumulation of individual experiences. If one thinks of the group as a machine that processes problems in an effort to arrive at solutions, the group process is the description of the interworkings of machine components, while the actual steps have to do with the function performed by each component. Group process issues are discussed at a conceptual level different from that of task issues. For example, while the group process issue of goal definition is usually foremost at the problem statement stage, it may also surface at the decision-making stage. Issues of group process are often associated with particular task steps but nevertheless span the range of problem solving, as well as being relevant when groups are resisting working. Figure 5.1 shows these interactions schematically.

The overt goal of the group process is to perform the group task. But groups by their nature have another, usually covert task: to achieve identity as a group. This is not to say that groups always work to perpetuate themselves (although many do) but that they strive to maintain a group identity and may do so in ways that interfere with the performance of the task. For example, if members fear that new ideas will upset a leader who has authority to disband the group, they will tend not to be innovative. In this way,

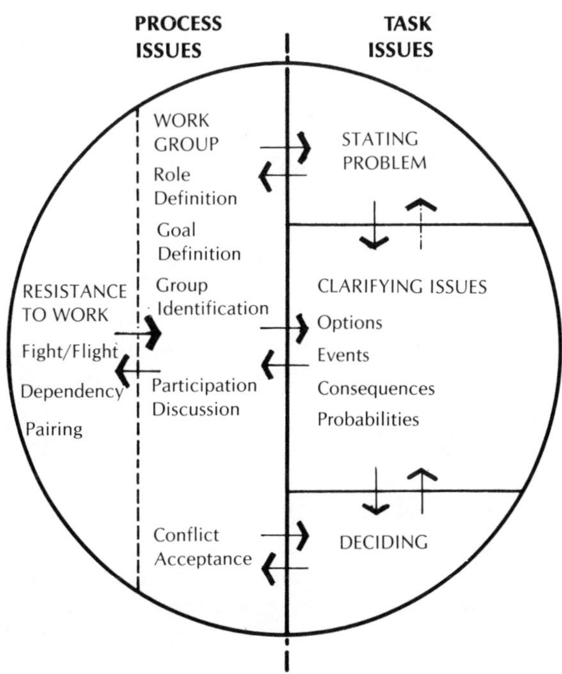

Figure 5.1 Group process and group task.

they try to promote the group's continuance at the expense of not producing new ideas. When a group is manifestly not carrying out its task efficiently, it is always to the point to wonder what group process goal it is trying to achieve. In well-functioning groups, the achievement of the task improves group morale, which in turn leads to improved group functioning and better task performance.

This chapter will take as its central question how to affect group process to improve small-group problem solving. Since the leader generally has more influence than any other member on group processes, most of the suggestions pertain to techniques the leader might use. Members may use many of these same techniques.

ESTABLISHING A GROUP

Why a Group?

A person (or another group) with a task to be completed must decide whether to use an individual problem solver or a group. Planners use groups largely to bring together varying expertise, to help accumulate data, to lend legitimacy to the decision-making process, and to provide a range of alternatives, options, and ways of thinking not available from one person. Not all decisions are put to a group because group work is time-consuming, someone in authority already knows what he wants to do or is satisfied that he can figure it out, or the problem solver is concerned that in trying to satisfy disparate group members, poor decisions will be made.

Plan for a Group

Once a task is defined, a group can be established. Its structure can be specified by discussing the nature of the leader, the members of the group, and the external factors relevant to the group. Most planning groups will have a leader specified. The leader may be appointed, may set up his own group, or may decide to form a group that will elect its own leader. In our example, Leader was designated by the consultant firm. It is a natural tendency of groups to want leaders; even so-called leaderless groups tend quickly to assign certain organizing functions to one person. To see how strongly a group will try to obtain a leader, one need only look at psychotherapy groups in which the therapist actively stays out of the leadership role. In such groups, one quickly sees anger toward the therapist for not fulfilling the group's expectations that he lead. Some groups rotate members through the leader's role, either because no member wants the continual chore of leading or because the group is afraid one leader will become too powerful. The potential difficulty with changing leaders is that a group will spend a large portion of its time reestablishing role definitions, expectations of how the group will operate, and modes of interaction.

Leader

A group's leader has a number of decisions to make in organizing the group. What kind of a leader is he going to be? This question can be asked along a number of dimensions. How much authority will he wield? Can he, as leader, constitute a majority of one and outvote the entire group? Does he see the group as advisory to him? Is he going to run a formal group according to Roberts' Rules of Order or allow a more informal exchange? Does he relish loud, heated discussions, or would he prefer courtesy and decorum to be maintained at all costs? Many of these decisions will be made without thinking and will seem to the leader to be determined by previous experience, outside expectations of the nature of the group, and the leader's personality. They are decisions that often have an "of course" quality, although groups are potentially capable of working effectively with widely varying leadership styles. Still, it is useful, particularly for beginning leaders, to be clear about these various dimensions because the answers here will affect the choice of members as well as giving the leader a clear expectation of what his role will be.

Regardless of the sort of authority the leader intends to wield, it is of crucial importance that he be clear about and identify with the group's task. If his leadership is seen as facilitating the work, the group will identify with him. In many planning groups, the leader has primary authority to make decisions. As far as the group goes, however, his major role in the group process is to facilitate the group's carrying out its task. This role is different from whatever power prerogatives he carries. A leader with authority must be both facilitator and decision maker. He will do better if he acts as a facilitator most of the time. Some leaders temperamentally have difficulty setting aside their authority to help to promote group discussion. Such people in authority might well ask another group member to act as discussion leader and facilitator so that the authority can participate actively in the discussion without having to be so concerned about the dampening effect his comments would have were he also the facilitator. Even as a group member, a person with authority needs to focus on facilitating the group and defer decision making until the group has made its opinions known. Groups do not work well when they feel their contribution is undervalued. If they are to meet simply to rubberstamp decisions the authority will make independently, their task, properly defined, is only to legitimize, rather than produce, solutions. The most common reason for unproductive meetings is that the leader fails to recognize that his primary responsibility is to help the group work on its task. This does not mean he forfeits his powers of decision making but rather that he suspends them until after full group discussion.

Because of his role as group facilitator, the leader must be prepared to suppress expressing his own feelings. Other members may show anger at a disruptive member, but for the leader such feelings are properly expressed only if they serve to help the group as a whole function effectively. The higher the leader's status in his organization relative to other group members, the more important it is that he remain impersonal in the group.

Membership

In the Water Group, Leader will pick the members. First, how many should be in the group? Groups larger than ten become unwieldy and more formal. If the number of experts needed is larger than ten, it is wise to set up several small groups. Groups of five or fewer are so small that the absence of one member can bring the group's progress to a halt.

Leader has some preliminary ideas about who should be in the group. Planner B, another member of the firm, has wide expertise in energy planning. Leader thinks that a planning engineer will be necessary. Both states are under the regional jurisdiction of the Western Natural Resources Board, and a representative from that board seems appropriate. The office of water ministry from each state should have a representative. Leader sees that there is room for several more people should he want them. Bringing in some private interests would expand the range of points of view.

Increased heterogeneity in a group generally provides more ideas at the expense of a greater level of conflict. Leader enjoys lively discussions, so he decides to have a representative of an environmental action group and a utility company representative. And he thinks that a planner whose expertise is financial management would be useful. The final list of roles on the committee is:

1. Leader.
2. Planner B (general expertise).
3. Planner C (engineer).
4. Planner (fiscal planner).
5. Representative of Natural Resources Board.
6. Representative of water ministry, Hiatonka.
7. Representative of water ministry, East Victoria.
8. Representative of environmental action group.
9. Representative of utility company.

Leader, having decided what roles he wants in the group, now chooses the particular people. In this task, Leader will be guided by several considerations. First, how well do members get along? In determining the other planners in the group, Leader has already decided on General, a planner with general expertise. He chooses Engineer because of his skills and Biz Whiz, a fiscal planner. Leader, having worked with the planners in his firm before, is confident they can all get along well. That members do not know each other is no reason not to include them in the group. In fact, if a significant subgroup is well acquainted outside the group, they may bring ways of interacting with one another that lead other members to feel excluded. This is a potential pitfall for the Water Group because the four planners are friends and have worked together many times before, so Leader plans to be attentive for coalitions.

Group cohesion increases with personal positive feelings between members. Leader might face a trade-off between, say, Mr. X of the Natural Resources Board, who has a great deal of expertise but an abrasive personality, and his assistant, Nat R, who has slightly less expertise but whom everyone likes. It is not simply a matter of personal preference to choose the likable Nat R; the quality of group performance will increase if the meetings are pleasant and cooperative. One hostile member can make work so unpleasant for the rest of the group that productivity falls drastically. Leader might arrange a side meeting among Mr. X, Nat R, and himself or utilize some other informal arrangement with Mr. X to make use of his expertise without including him in the group. The water ministries are picking their own representatives, Hiatonka and Victoria, so Leader has no choice there. The environmental action group frequently sends Mr. Slowfeet to planning meetings. Leader has reservations, not because Mr. Slowfeet is not bright but he is often gone, and when he comes, he comes late. Groups tend to regard such behavior as a member's voting with his feet on the group's worth, with a consequent lowering of morale. Leader wants the group to go well and so invites Environ, who accepts the appointment. Mr. Noisome, the in-house planner for the utility company, is known for his loud blustering at everyone who disagrees with him, so Leader invites his assistant, the equally competent but more temperate Utility. The Water Group membership is therefore as follows:

1. Leader (Planner).
2. General (Planner).
3. Engineer (Planner).
4. Biz Whiz (Fiscal Planner).
5. Nat Resources (Representing natural resources board).
6. Hiatonka (Representing water ministry).
7. Victoria (Representing water ministry).
8. Environ (Representing environmental action group).
9. Utility (Representing utility company).

External factors

Specifying the members and the leader demarcates only one side of the group's boundary. External factors, those persons or agencies not in the group but relevant to it, demarcate the other side. The planning consultant firm that set up the group and defined its task is one such factor. The directors of the departments of energy, to whom the final report will go, are others. Because the nature of planners' work affects outside groups, planning groups must take cognizance of such external factors. All members of the Water Group serve not only as members but also as liaisons to outside groups. Since planning groups often take part in preparing reports and recommendations for other groups, ascertaining the activities, expectations, and preferences of such groups is essential. Planning groups often use members as representatives of their constituent groups, both to obtain varying points of view and to legitimize group decisions. It is important that members be selected whom the constituent groups will acknowledge as representative. For example, the Water Group might use the membership of Utility to rebut a charge that their final report did not properly consider the utility company's position. If the utility company did not have faith in Utility's ability, his membership could not be used in this way.

External factors can operate to the detriment of the group. Political intrigue in the consulting firm may lead to the group's disbanding. The Water Group may fear that the energy departments will give its report little study regardless of its content. Such difficulties interfere profoundly with a group's work. In such cases, the group may need to address itself to the external factors to ensure its own survival and usefulness. On the other hand, anticipating that a report is eagerly awaited and that a great deal turns on it can help to bring meaning to the group's work, increase the group's cohesion, and improve dedication to the task.

THE FIRST MEETING

The Water Group thus far exists only in the minds of the leader and members. At the first meeting it becomes an active group. The group has two jobs. The first, and obvious one, is to work on its task. The second, less obvious but in some ways more powerful, is to keep itself going and maintain a group identity. A group identity must be formed to some degree before the group can begin work on its task. The identity will continue to develop over the life of the group, but initially the group needs to share a sense of roles, power, and the task at hand. This process often takes place without being noticed, but it becomes highlighted when consensus breaks down. The structure of the group is like eyeglasses: most of the time the user is not aware of his glasses, although they greatly affect what he sees. He becomes aware of them when they fall off, or get dirty, and the function they perform is not being carried out. Structure is not noticed as long as it is shared. As the group goes along, it may become evident that certain elements of structure that were thought to be shared are not. Suppose at the first meeting Utility does not recognize that Leader should lead and insists the group elect its own leader. The group has several options: to follow his idea, to reject it, to table it, or to vote Utility out of the group, but they cannot easily ignore his statement. A challenge to the existing structure must be addressed before work can proceed.

Building Group Identity

Making introductions

The group's first meeting has great influence on setting the tone for meetings that follow. The group probably will start by making introductions, an ideal time for the leader to help members to get to know each other. Instead of having each member just state name and title, the leader can say why he wanted the person in the group, and go on to ask the member to expand on his background, tell the group why he is there, and say what he hopes to get out of being a member. This introduction helps the group to get to know each member and encourages members to begin to define their own roles in the group. After introductions are made, the leader may wish to make a few further clarifying comments about how he anticipates the group will work.

Recorder

Most groups should have a recorder who will summarize the group's major points of discussion and distribute notes by the next meeting. Notes allow members to refresh their memories about the previous meeting. They also, and perhaps more importantly, acknowledge members' contributions and let members know whether they were properly understood. Groups differ in their need for formal documentation to present later to clients, other planning groups, or outside agencies. Some members may want only decisions recorded, thinking discussion will be freer if it is off the record, while others may wish for more detail. The first meeting is a good time to raise this issue explicitly for group discussion.

Agenda

Agendas are useful for small work groups, especially at early meetings. They help to give a sense of direction and set goals for the meeting. By identifying goals, agendas can help forestall getting stuck on one issue. The agendas may not need to be distributed to each member but could be put on a chalkboard. At an early meeting, a group should consider its use of agendas. Some leaders prefer to have a formal agenda that must be adhered to and use it to control meetings. Other groups use agendas as skeletons for structuring meetings and are open to additions during meetings. Informal groups usually do better when members feel free to add items to the agenda. Agendas for the first meetings often encompass such issues as setting priorities and directions for later meetings. As it develops, the group should also consider its longer agenda, the overall flow of its work, in order to fit its work efficiently to external time constraints.

Setting Forth of Group Task

Leader now, relatively briefly, sets forth the group task. Assuming he has made extended introductions, he has some sense of how the group task fits in with members' expectations. In specifying the task, he will also want to set down relevant external parameters such as how much time the group has to prepare its report, how often and where it will meet, whether funds are available for doing studies, and how long the report is expected to be. After briefly discussing the task, he should encourage the group to respond.

During the introductions, Engineer says he has been on numerous groups writing reports of this sort before. "Speaking as a planner," says he, "I know that many reports get filed in the circular file. I'd be a lot happier if I knew who was going to read our report." When Leader then specifies the group task, he can take particular care to identify who will be reading the report and in the ensuing discussion should check with Engineer to see whether he still has concerns about this.

Speaking to members' concerns and then asking for their response is useful even if the information does not allay the concerns. If members feel understood, they feel an affinity with the group and its task even if they do not get what they want. This principle is of central importance in group decision making because it allows members to compromise their individual preferences in favor of a group decision. If Engineer were dissatisfied with Leader's answer, Leader could then take the next step and note that to whom the report goes should be considered a problem with various options to be explored. This underlines that Engineer's contribution has been heard and may in fact lead to better use of the group's report.

In this first meeting, members should have time to react to each step. The group will likely see some testing between members and subtle jockeying for roles, position, and influence. The less familiar members are with one another, the more time needs to be allowed for this initial testing. Allowing the group process to develop in this way does not mean delaying the task but recognizes that a group needs to establish an identity before it can work. At the same time, group identity grows stronger when a group does meaningful work. After members have introduced themselves, explained their goals, discussed the group goal, and appointed a recorder, the group will move to problem solving.

FACILITATING PROBLEM SOLVING

Using the problem-solving schema introduced earlier, we will use two examples to illustrate how group process affects the performance of problem-solving steps. We will continue on with the Water Group's first meeting and follow with a more technical example.

A Process Problem: What Should the Group Do Next?

Stating the problem

Leader would like each member to bring to the next meeting a three- to five-page memo of the relevant concerns in that member's area and considers simply asking each member to do this. Such a directive runs the risk of causing the members to feel ordered about, incapable of deciding, and devalued. An important principle in working with groups is to try to turn individual decisions into group decisions, so Leader asks the group specifically how it should proceed.

Clarifying options

At this early stage in the group, Leader is particularly attentive to setting the tone of discussions that will follow.

Promoting informality Small groups tend to work best if they proceed informally. Using first names, for example, helps promote free and open discussion. The leader will want to reduce inhibitions so as to allow the personal attractions between members to increase. Reinforcing members for their contributions helps to promote informality and reduces anxiety about contributing. The leader should recognize that groups generally wish to promote harmony and positive group feeling and so are often quite responsive and ready to accept initial efforts in this direction. In informal groups, shifts to formal procedures, such as a move to parliamentary procedure, generally indicate the existence of unresolved conflict.

Encouraging participation For groups to work well, it is important that all members participate. Promoting informality is intended to encourage participation. Members will participate when they feel their contributions are valued by the group. Everyone has had the experience of saying something and having no one respond—one feels rather like a cartoon character who runs straight off a cliff and is sustained in the air until he realizes he is no longer on solid ground. Reactions to members' comments, such as overtly appreciating them, restating them, clarifying them, and integrating them with what others have said, all promote participation. For members who hold back making comments, active solicitation of their ideas by the leader is often helpful. One easy and useful method for integrating ideas is to write them on a chalkboard.

Leader asks for suggestions and writes the following on the blackboard in response to members' comments.

1. We don't need any hydroelectric power; people could conserve.
2. Do a stream-flow analysis to ascertain whether the Pleasant River has enough energy to be worth damming.
3. Prepare three- to five-page memos.
4. Begin a more detailed site analysis of Eagle Point (in response to Nat R who thinks from his own work that this is where a dam should probably go).

Leader writes these suggestions on the board without criticism and without specifying names. This serves to define the ideas as group ideas, and so, when one is turned down, it is the group that is discarding one of its own ideas rather than rejecting a particular member's contribution.

Clarifying consequences of options The group then turns to look at the implications of each suggestion.

1. Environ's suggestion about conservation would lead the group to disband. A member comments that this suggestion should be redefined as describing a trade-off problem, and Leader comments this will need to be addressed later.
2. A possible consequence of stream-flow analysis is that no dam may be feasible.
3. Memos (Leader's hoped-for result) would allow the group to begin exploring a wide range of issues.

It often happens that a group feeling positive about itself will set goals beyond its capabilities.

Engineer, having just completed a report on dams in the two states, says he will prepare an abstract of dam usage for next week's meeting. Hiatonka and Victoria, not to be outdone, promise to have ready a detailed report of all water usage for energy in their respective states. General, having recently worked with the state water ministries, fears that a complete survey of current usage cannot be completed in a week and is concerned about the effect on the group if the first deadline it sets for itself is not achieved. He proposes, therefore, that the three surveyors begin to collect relevant data and report their preliminary impressions to the group at the next meeting. Leaving the performance of the first task somewhat open-ended allows the members to try to achieve what they had initially stated but does not carry the penalty of failure should that project prove too ambitious.

4. The consequences of a site analysis would be high cost and significant time required.

Choosing among alternatives

Having clearly specified the consequences of the four options, the group rapidly reaches the consensus that a stream-flow analysis has to be done. They also think short reports are useful and decide to table the site analysis pending further analysis and to delay discussions of the value trade-offs inherent in the suggestion about conservation. Had Leader overlooked or been unaware of the need for a stream analysis or had he specified that other members should bring in reports, he would have at least delayed getting the stream-analysis data and run the potential risk of the group's proceeding without that crucial information.

A Technical Problem: Effects of a Dam at Eagle Point

Problem statement

Several meetings later the group has located various possible sites for dams and is now considering Eagle Point. The members seem agreed that they need to give this location special attention, and the problem has become one of determining the consequences of putting a dam at Eagle Point. The problem statement also includes a specification of likely outcome criteria such as power generated, cost, and environmental impact.

Indian ruins that would be flooded by a dam, with major implications for overall feasibility. Nat R raises the possibility that the need to retain water in a reservoir behind the dam would limit generating power. The Water Group cannot anticipate all such problems but will have a much better chance of finding them if it occurs to someone to look.

Option clarification, external events, and judging consequences

During this phase of problem solving, the group collects and weighs information. They set out the possibilities before arriving at a decision. The greatest danger for the group in this phase of problem solving is to reach early closure and move into the phase of making choices without fully setting out all relevant parameters. One of the greatest advantages of groups is the potential for generating diverse ideas; one of the greatest problems is that the group process might prevent this from occurring.

Groupthink I. Janis coined the term *groupthink* (Janis, 1972) for the phenomenon that occurs when groups become so harmonious that diversity is lost and new ideas do not surface. This can occur when groups seek conformity, when conflict is seen as too dangerous, or for other reasons. The leader's primary job in this phase of problem solving is to prevent premature closure.

The Water Group had listed four topics—engineering questions, environmental impact, cost, and community relations—and appears ready to move on to discussing other sites. Leader suggests that the group consider further and in more detail possible adverse consequences of a dam at Eagle Point. The group lists a number of possible consequences, and members start mentioning other dam locations. Leader, concerned about premature closure, asks, "What might we have left out?" As discussion proceeds, Engineer, who was saying that a detailed flow study should be done (at significant cost), is reminded that the Army Corps of Engineers might have done a low-level engineering study; if it had, the corps' report would save a great deal of the group's time. Environ questions whether there might be an archeologist studying

In addition to keeping open the search for more options, Leader has some particular techniques for promoting a range of ideas.

Brainstorming Brainstorming is a technique that generates new ideas by expressly forbidding any evaluative comments of ideas until all ideas are listed. It is often conducted by asking members for ideas and listing them on a chalkboard as they talk. Seeing others' ideas, members are encouraged to respond however they wish, and these remarks are listed without evaluative comments. In groups where some members are much more verbal than others, a further way of promoting ideas is to have members write them down anonymously. This prevents less assertive members from being interrupted and adds anonymity to the safety of no evaluation. Notes are collected, the ideas listed, and another round of idea generating is conducted. The primary difficulty of brainstorming is that it seems gimmicky in situations in which members do not fear others' comments.

Devil's advocate The leader challenges group thinking as a matter of principle, not necessarily because he believes in or is committed to the position he is espousing.

Evaluation of progress thus far Often the group can look back on its prior decisions and try to find problems with them, and then consider whether these problems are fully accounted for in the current list of options.

Conflict By bringing together a diversity of views, groups can obtain a synthesis that a single person could not produce. Conflict is useful in helping to bring these diverse views forward. As members marshal their arguments, new and varied points will be raised. The danger is that conflict will become so heated that productive work ceases. Members will begin to use their ideas as supportive arguments rather than for illumination.

The resolution of unproductive conflict is a major concern of the group. Generally conflict can be tolerated to the extent that it does not threaten the group's identity. For this reason, early sessions are much more sensitive to disruption by conflict than later ones.

At the first meeting, Environ and Utility, having battled elsewhere, begin to discuss rather heatedly the trade-offs between environmental protection and resource utilization. Leader, wanting to build group identity, notes that there is clearly disagreement, summarizes the issues raised by the two members, and comments that the group will need to work through this conflict at some point, but comments he would now like to move to the next point on the agenda.

Leader's intervention recognizes the existence of conflict, affirms the importance of the positions by commenting that the group will take up the issues, but delays the discussion until the group is more cohesive.

When conflict erupts, the leader will serve a number of managing functions. The primary goal will be to keep to the subject at hand. The leader wants to discourage personal attacks, to integrate the material with earlier group work, and to defuse emotions if the discussion becomes heated. Leader accepts the presence of conflict but attempts to limit its scope to useful discussion.

As the group is discussing the pros and cons of a dam at Eagle Point, Environ and Utility take up the discussion they left in the first session. Utility begins, stating, "All the environmentalists want to do is lie down in front of the bulldozers." As Environ starts to redden, Leader asks Utility why he thinks the environmentalists might do this in the specific case of the Pleasant River Dam. This comment defines the issue as task related and prompts Utility to role play the environmental position. Before Utility can respond, Environ remarks, "People like Utility dream with glee of driving bulldozers into mountains of environmentalists."

At this point Leader has a choice: He can try to stay in a problem-solving mode, or he can decide that the personal animosity needs to be addressed immediately. The group cannot well handle both issues at once. Let us suppose he decides the personal animosity is not great, that the two group members are taking up established rhetorical positions, and that the feelings involved can be discounted for the present.

Leader says Environ does not know Utility's dreams. This defines Environ's remark as inappropriate. Leader then returns to Utility with another request for his understanding of the environmental objection to the Eagle Point Dam. Utility says he thinks environmentalists want to canoe on the Pleasant River. Environ responds that that is only one of the difficulties and lists three others. Leader then turns to Engineer, who has recently completed a dam survey, to see what effect earlier dams had on recreational use, which integrates this discussion with an earlier issue. After getting a response, Leader summarizes the sense of Environ's objections to the dam, and invites Environ to discuss with her environmental action group which of these issues are most relevant, thus identifying the next procedural stage for resolution of this conflict.

Conflict at an emotional level is handled somewhat differently. Ignoring or discounting it seldom leads to resolution. The group may sense that the conflict is too disruptive and shy away from it or may feel that a good statement of feelings would be useful. If we go back to our example, after Environ makes her comment about Utility's dream, Leader may choose to involve the group in the decision about which level to work on, saying, "There seem to be some strong feelings involved here. I wonder if we need to get them out in the open before going on," and then waiting for the group response. If the group moves to discuss feelings, it leaves the level of problem solving on the issue of the dam and moves to the level of group process. The central principle in dealing with conflicting feelings is that feelings should be accepted, not criticized or externalized. Thus, if Environ goes on to say, "All

Utility does is bicker, put down environmentalists, and want to get his own way," the important thing to focus on is Environ's feelings and their clarification, not whether Utility is really that particular way or not. A group member might say, "And you get angry when you think Utility is out to provoke you." The focus remains on Environ, not on Utility. Utility may chime in, with a denial or acknowledgment, and the group can then take the same approach with him: "You feel strongly about this" or "Did Environ's response to what you said surprise you?" The leader should avoid taking sides. Some groups will try to scapegoat one member, and it is then quite important for the leader to support the scapegoated member without attacking the group. Generally people feel better after they have expressed their feelings and felt someone else has understood, even if the external reality is unchanged. If the group continually needs to discuss these kinds of conflicts, it should try to identify why these feelings continually emerge.

Conflict between a member and the leader has special characteristics. As a member of the group, the leader may want to speak his mind, but he must realize that he is also the leader, and if he abandons the principles of keeping conflict impersonal and encouraging group discussion even when the group disagrees with him, he will have a major adverse impact on the group. He needs to clarify and promote full discussion in his role of group facilitator, at the cost of not exercising some of the privileges of group membership.

The examples of conflict presented so far concern two people. Conflicts sometimes arise among three or more, and the general principles for handling them are the same. Disruptive conflicts tend to reduce to a succession of two-sided conflicts, say with A, B, and C all disagreeing, leading to A and B against C and then A against B.

Group responsibility for group process In the examples thus far, Leader has taken the primary responsibility for managing the group process. In well-run groups, not only the leader but each member learns to share responsibility for the groups running smoothly. A member learns by watching the leader and modeling themselves on his ways of facilitating the group's work. For example, when the leader says, "I don't want to take sides," as he responds to a conflict, he makes his technique explicit. The leader can encourage members to facilitate the group by commenting positively on their actions when they promote the group process. The leader may also ask the group at the end of the meeting how they thought the meeting went, and if there were difficulties, how the group might better handle such problems. This last technique may seem gimmicky with some groups, particularly those that have worked together for some time.

Victoria begins the meeting by reporting that Leader called her earlier to say he was ill and could not attend today's meeting. The group appoints a temporary chairperson and goes on with their discussion of the Pleasant River Dam. As the discussion seems to come to a premature closure, Victoria takes a devil's advocate position. This leads to a heated conflict, which Biz Whiz defuses by restating the work-related issues involved.

The members are able to take over Leader's functions in his absence, and perhaps they can also carry out many of those same functions while Leader is present.

Choosing among alternatives

After clarifying options and consequences, the group comes to the point of decision. Planning groups frequently do not make final decisions on issues requiring value trade-offs and consequently are prone to feel that their conclusions carry little weight. For instance, they will not make a final decision about whether the power a dam at Eagle Point could generate is worth the negative environmental impact and expense of constructing a dam. Such decisions are usually made in the political arena.

Groups can be helped to be more comfortable in their advisory role by fully discussing the decision-making mechanism both in and out of the group. The task of an advisory group is to decide in specific terms what recommendations they are making to the final decision maker. To maximize the usefulness of the report, it is important for the group to understand the process by which a decision will be made after the report is sent. From a group process perspective, such discussion helps clarify the external factors of the group's structure. This can help the group to formulate recommendations in a way that will directly address the concerns of the decision maker. Some group members will have other relationships with the decision maker and will be in a position to undertake some political activity after the report is sent. For example, in the Water Group, Hiatonka and Victoria work with the directors of the departments of energy and so both can contribute useful information about how their departments use consultants' reports and also can discuss the report with the decision maker after it is sent.

Most groups wish for spontaneous unanimity in the decisions and, failing that, will try to arrive at a compromise consensus. This tendency derives from the group's wish to have members think alike and is

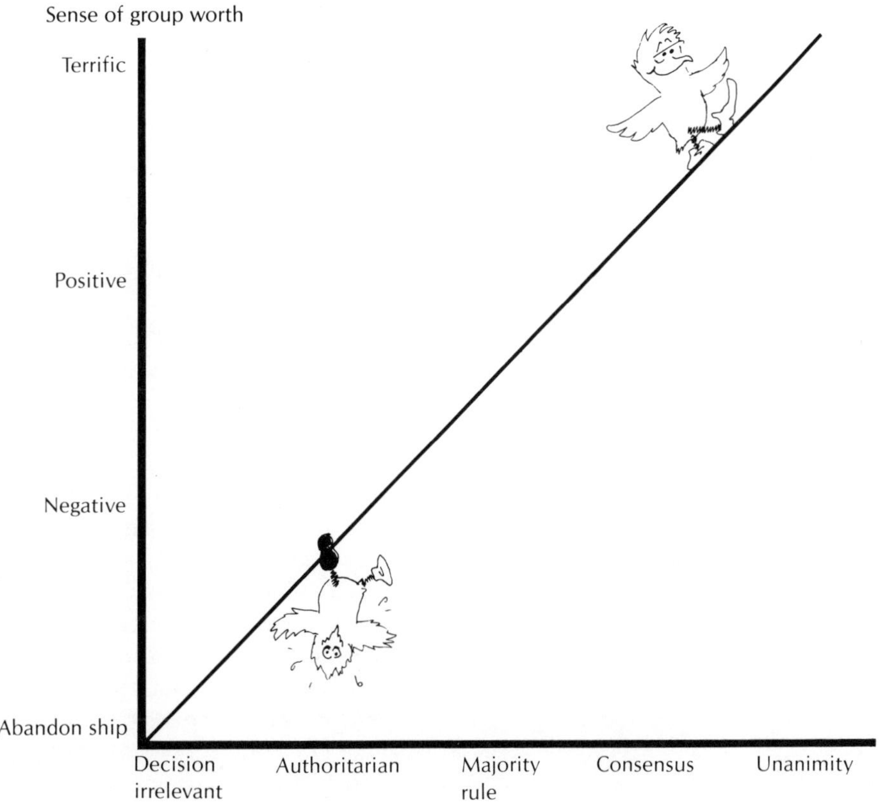

Relations between feelings of group worth and decision mechanism

closely tied to the group's wish for cohesion. Since a decision not accepted by all members highlights that views are not shared, groups tend to avoid making these differences manifest out of concern that those with different ideas will feel excluded. Groups can tolerate some decisions in the face of continuing disagreement, but if this becomes frequent and the opposition vocal, the group's identity tends to be threatened. If the group does not agree, a useful first step is to elucidate those issues on which the group can agree. Often a group can agree on technical assessments but not on value trade-offs.

Environ and Utility both read the ornithology reports and agree that a high head dam would likely lead to a 40 percent reduction in the number of eagles nesting at Eagle Point. Utility is not bothered by this, while Environ is quite upset and mysteriously predicts, "God gets even!"

Even if the group knows that it will not be making the final decision about whether to build a dam, it can still emphasize or deemphasize the importance of a 40 percent reduction of eagle nests. The next step is to air fully the group's concerns about the reduction. Often if members believe they have had a chance to make their points, they can accede to the consensus, perhaps by stating, "But the report should have a few lines about the eagles' plight," and they may even then go on to support the idea to their constituent groups with, "It's not so bad—at least we prevented wiping out the fiddlefish." Members do not come to the group expecting to get everything they want. What is important is that they feel that the decision process took them into account. In planning groups a minority that feels strongly is usually not outvoted but allowed to submit a minority report.

GROUP RESISTANCE

Leader is becoming uncomfortable with the group. Everyone is friendly, cheerful, cooperative, and kind, but with something of a jolt, Leader realizes that for the past several weeks no progress has been made. Leader thinks back. At the last meeting the group had a long discussion about how the fuel industry lobbies were working to get their oil deregulation bill through Congress. The group had much imaginative discussion about how they could stymie the lobbyists if only they were the congressional subcommittee acting on the bill. Since they were not, the discussion led to no useful work. After that protracted discussion, the group moved on to gossip about the personal life of the governor of East Victoria. It was quite an absorbing discussion but not useful. Previous meetings had been similar. What was wrong?

When groups move away from their task, they have left the work group mode and moved to another level of functioning. The first step— often the most difficult—is to recognize that this has happened. The difficulty stems from the level of absorption that can be present in the group while discussing matters not related to the task. Often the recognition that the group has made such a shift will come from a sense that nothing is being accomplished although the discussion *seems* relevant. Once this is recognized, the leader or any group member can move to a different point of view: the perspective of group process.

When groups are functioning well, the group process is not at the forefront of attention. Instead the group is working with the issues that need to be resolved to accomplish the task. Looking at the group process arises out of a need to account for and deal with resistance to working. This shift in perspective can be

Figure 5.2 On perception (after Boring, 1930).

likened to a visual shift in perspective. Those not familiar with the illustration in Figure 5.2 will, at first glance, often see a young woman. If told that there is another way to see the picture, some people can find it alone; for others it helps to know that there is a picture of an old woman in profile. In groups, to see the group process, one needs to shift away from the more usual level of focusing on individual interactions.

Types of Resistance

Many different perspectives of group process have been delineated by differing theorists. W. R. Bion, from his work with psychotherapy groups, has set forth one way of conceptualizing group process (Bion, 1974). Bion speaks of groups, when they are not working, as having a purpose other than solving the group task. This purpose is usually not conscious. Bion delineates three purposes, which he calls basic assumptions, as the most common.

Fight/flight group

The group is gossiping about the fuel lobbyists. General bemoans their pernicious influence on legislation. Biz Whiz mutters about possible bribery. Environ is sure they will work to suppress the group's report. Engineer relates some juicy tidbits of one lobbyist's recent divorce.

The group is eagerly talking about the fuel lobbyists, focusing on "those baddies out there" in a way which uses the group's time but accomplishes little. Bion refers to a group operating in this way as having the basic assumption of fight/flight because the group views its responses as either to destroy the danger or to flee it. These responses may be couched in terms of wishes or intentions rather than planned actions. Such a group needs to be distinguished from a working group discussing, for example, what review process the group's report will undergo, perhaps considering the possibility that lobbyists might affect that review. A group can move to a fight/flight mode and spend hours bemoaning the inadequacies of the reviewers and fantasize about ways to get even with them for their supposed incompetence, rather than discuss in a work-oriented mode what factors about the reviewing committee need to be considered. One of the signals that a group is in a fight/flight mode is that members talk of the group as though it is the home of all things good, while the external force or agency is seen as the embodiment of badness, devoid of any redeeming qualities. The group, however, can itself achieve cohesion in the process of focusing on a perceived external danger. At times the group's need for an external danger is so strong that it will create one in the group mind even if none exists. Presumably this is useful for the Water Group although why the group needs to do this is not yet clear.

At the next meeting the group again picks up its discussion of the lobbyists. Leader says this discussion is interesting, but he does not see how it directly affects what the group needs to do next, which is to assess the effects of damming the Pleasant River. The group agrees, with some chagrin, that he is right.

Leader has clarified the basic assumption of fight/flight, pointing out its essential unreality, and this allows the group to change.

Dependency group

After a pause, Leader suggests the group take up the next issue on the agenda, preparing the assessment of the dam. Nat R says he does not know what form the report should be in and looks to the rest of the group for help. Other members also seem to have no idea what form would be most appropriate, and one by one they suggest Leader should guide them. Leader, attempting to be helpful, raises several possibilities, all of which the group finds faulty. The group seems agreed that it cannot solve this problem and needs direct instructions from Leader. Leader begins to have the feeling that he is leading a group of helpless children.

Such a group is an example of what Bion calls a dependency group. Dependency groups often depend on the leader although a group may see itself as dependent on an outside force. When Utility says, "We can't write the report until the department of energy tells us how," even though it should be clear that the department has less idea of what the relevant factors are than the Water Group, then the group is acting as a dependency group. Leader says the group seems to expect him to do all the work and points out that General, Engineer, Nat Resources, and Utility have all worked on similar projects before and probably have a good deal of experience in writing these reports. Everyone agrees, and members begin to talk about what formats have been used in the past.

Pairing group

After some profitable discussion, Environ and Utility get into a long, repetitive argument about the effects of damming the Pleasant River. The group has heard this many times before, and little new is being said, but the group nevertheless patiently sits back and listens as this discussion goes on and on. Twenty minutes later, it is still going on, with no end in sight. Environ and Utility are discussing the issue more heatedly, although what is being said is repetitive. The rest of the group looks intensely interested, although exactly in what is not clear.

Bion refers to this phenomenon—two people taking center stage as the rest of the group sits back and watches—as a pairing group. The basic assumption in a pairing group is that two people have taken over the group's functions; the rest of the group acts as though they were waiting for something productive to come out of the interaction, although there is little evidence that this will occur. The group fantasy is nevertheless quite strong and serves to hold other members back from participating. Leader, recognizing what is going on, says that he thinks the group as a whole should express its comments on what format to use and what considerations about the dam are relevant.

Working with Group Resistance

In these examples Leader is like a physician who makes a diagnosis and intervenes in a way that reduces symptoms, but the symptoms are then replaced by another difficulty. Leader's interventions help to get the group back on the track for a time, but as he goes home, he has the uncomfortable sense that something still is not quite right. After each intervention, the group shifted into some other form of resistance. Why?

The group's shifting away from a work group to a fight/flight, dependency, or pairing group, represents resistance to the task at hand. The causes of such resistance are not always easy to find, but it is useful to bring them into the open. Repeated moves out of the work group state usually indicate that some feelings are being avoided. The work group leader, considering the problem, decides to take action.

At the next group meeting Leader finds that the group within five minutes is having another jovial discussion about the governor of East Victoria. Not knowing whether this is social pleasantry, Leader waits five minutes and finds the group still on the topic. He then verbalizes his concern that the group is having a problem accomplishing much, not only now, but over the past three meetings, and suggests that something is going on in the group to cause this, but he does not know what. Members, on reflection, agree. After a pause, Utility says he finds himself a bit confused at the meetings since he does not get agendas ahead of time. An underlying theme in his tone of voice says he is resentful that the group does not think enough of him to get the agendas to him in time. Environ says she is annoyed that Biz Whiz does not come to meetings but is still a member and feels tired of always waiting for her. Leader suggests that maybe something is going on in the group Biz Whiz is staying away from. After a lull, Nat remarks that several meetings ago he submitted a preliminary report that the group had quietly, but strongly, criticized. He said he felt all his effort had gone unappreciated. General says he tossed out what he felt were some good ideas and no one seemed to care. Leader says he had been concerned when he heard the head of the consultant firm say, "Who cares about the damn water anyway?" When a member voices the idea that individually each member was feeling unappreciated, there is a general nodding of heads. Engineer then points out that the group is on schedule with its preliminary assessments, that he personally likes all the other members, and that he wants the group to do a good job. Other members begin to cite the group's accomplishments and start to evince positive attitudes toward continuing.

In this example, Leader keeps inviting the group to reflect on feelings that might be getting in the way of useful work. After the members are able to voice feeling unappreciated and isolated, the group is able to move attention to its successes, reconstitute the collective group self-esteem, and move on with some hope to productive work.

PROBLEM PERSONALITIES

Groups tend to blame difficulties on individuals. Such a view does not consider that what looks like problematic individual behavior may be symptomatic of an assumption the group is making. When Environ and Utility are arguing, it can appear as though two people are in conflict. Such a view neglects to consider the role of other group members who are allowing the argument to go on. One may see group involvement more directly by picking up instances of covert encouragement.

During a lull in Utility and Environ's argument, Engineer mildly says he supposes Environ is not understanding Utility's point, when it is clear that Environ had understood but simply disagreed. Utility accepts Engineer's statement and begins to recite again his views for the alleged benefit of Environ, who already understands perfectly well.

Here Engineer as a representative of the group can be seen to facilitate the continuation of a pointless and repetitive argument. Victoria, who apprehends the group process, says she thinks the group understands Utility's view and notes some members disagree with it.

At times, however, disruptive individual behavior needs to be handled directly. These problems occur particularly in groups that are afflicted with members whose personalities grate on the collective group nerve.

Hiatonka begins continually to interrupt the statements of others, makes personal and derogatory comments, challenges Leader's authority, and skips meetings. The group becomes increasingly resentful.

Or consider Victoria who likes to pick fights whenever she can. Or Biz Whiz, who is chronically late to all appointments, the Water Group included. Leader, who knows these three socially, knows their behavior is not specific to the group; they always act that way.

Fortunately groups exert great power over their members, and a cohesive group can often manage problem personalities as a matter of course. Everyone has had the experience of saying something and getting no response. The speaker usually has a sense of not being there and then is hesitant to say anything further. People who are difficult in their interpersonal relationships tend to require a counterpoint for their actions, and groups are capable of avoiding that reciprocal role. By focusing on the group's task, the leader can sometimes steer clear of the pitfall of pointless haggling. If the group process does not curb the problematic behavior, the leader has three options: to focus the group on controlling the difficult member, to focus on the interpersonal conflict in the group, or to confront the member individually. The range of maneuvers in these options is endless and much depends on the leader's personal style. It is helpful for all planners who work in small groups to become aware of their own coping styles. While many people think direct individual confrontations are powerful interventions, in fact the most powerful interventions are those that mobilize group pressure. Among these are statements that discourage what the problem person is saying or disallow it as a relevant contribution. As an extreme example, groups of psychiatric patients with a psychotic member can often stop that member from acting crazy in the group when any psychotic statement is defined by the therapist as irrelevant to what the group is talking about. The group gives the message, "If you're going to be like that, we'll have nothing to do with you." A threat of exclusion from the group, expressed or implied, has powerful effects on behavior. If the leader's statement does not suffice, he may encourage the group to confront the particular member about his way of relating. Usually if severely disruptive behavior continues in the face of confrontation by the group, the group can be encouraged to ostracize the member. Difficulties such as chronic lateness or absence often can be controlled when the group discusses their disruptive effects in the presence of the offending member. An interesting problem arises when one member is relatively weak and is not participating fully in the group's work. The usual response of groups is to expect less of that member and not push for that person's working up to the group's standards. Groups tend to be more tolerant of aberrant behavior than the leader or another individual member might be.

A group handles an instance of interpersonal conflict generated by a problem person much the

same way it handles other instances of interpersonal conflict, although with problem members, such conflict is more frequent. At an individual level, the leader or another member may meet with the problem member outside the group to discuss the difficult behavior. This produces less embarrassment than confrontation by the entire group. It is often less effective but can be a useful first step. Group confrontation can always follow.

TERMINATION

Most planning groups end after they have finished their task. In an effective and cohesive group, a positive group spirit and positive individual personal relationships will have been formed. The dissolution of the group will therefore be attended by feelings of sadness and loss, as well as feelings of satisfaction for a job well done. It is useful to allow time for the expression of feelings about termination and give members a chance to review what has been accomplished, comment on the successes and failures of the group, and to say good-bye to one another.

SUMMARY

Small groups perform specific tasks in planning. The leader has a particularly crucial role to play in shaping the group and the way it works, in structuring the group, formulating problems, leading discussion, resolving conflict, and facilitating decisions. When group resistance against working becomes manifest, shifting attention from the task to the group process itself offers the best way of dealing with the resistance. Understanding group process is a potent tool not only for working on tasks but for solving interpersonal difficulties that arise in the course of a group's work.

FOR FURTHER CONSIDERATION

Think of a small group of which you have been a leader or a member.

1. How would you characterize that group's structure? What was its task?
2. What issues of group process were problems for the group?
3. Recall several instances of conflict. How were they resolved? How might they have been better resolved?
4. What ways of encouraging discussion were particularly effective in that group? What ways did not lead to more discussion?
5. Can you recall discussions in which the group left the work group state and became a fight/flight, dependency, or pairing group? What was the transition out of such a state?

REFERENCES

Bion, W. R., 1974, *Experiences in Groups,* Ballantine, New York.

Boring, E. G.,1930, "Apparatus Notes: A New Ambiguous Figure," *American Journal of Psychology* **42**:444-445.

Janis, I., 1972, *Victims of Groupthink,* Houghton Mifflin, Boston.

BIBLIOGRAPHY

Berne, E., *The Structure and Dynamics of Organizations and Groups,* Ballantine, New York, 1963.

A readable discussion of the structure of groups of all sizes.

Bion, W. R., *Experiences in Groups,* Ballantine, New York, 1974.

The classic exposition of basic assumption groups derived from Bion's work with psychotherapy groups. Psychotherapists have studied groups in which the group task is to facilitate personal change and understand group processes. Such work tends to emphasize emotional issues, often with elucidation of the individual's unconscious processes and fantasies.

Freud, S., *Group Psychology and the Analysis of the Ego,* trans. James Strachey, International Psycho-Analytical Press, London and Vienna, 1922.

The original psychoanalytic theory of groups.

Hare, A. P., Handbook of Small Group Research, 2d ed., Free Press, New York, 1976.

Social psychologists have studied small-group behavior using formal research methods, using experimental groups under controlled conditions, and observing naturally occurring groups. Theories of social psychologists tend to be couched in terms of measurable variables of observable interpersonal interactions. The *Handbook* provides an extensive review of social-psychological small-group research done through 1972.

Harnack, R. V., T. B. Fest, and B. S. Jones, *Group Discussion: Theory and Technique,* 2d ed., Prentice-Hall, Englewood Cliffs, N.J., 1977.

A practically oriented, technique-filled book for working with small groups.

Janis, I. L., and L. Mann, *Decision Making: A Psychological Analysis of Conflict, Choice, and Commitment,* Free Press, New York, 1977.

Discusses the psychological resistances to good decision making. Uses some well-known and interesting examples of decisions that resulted in disasters.

Keeney, R. L., and H. Raiffa, *Decisions with Multiple Objectives: Preferences and Value Trade-offs,* John Wiley, New York, 1976.

A well-written and in part mathematical treatise on making trade-offs between different categories of values. Although the book uses some complex mathematics, the introductory parts of the chapters contain much of value for nonmathematical readers. Of particular interest to planners are several case examples from urban planning, such as the siting of the Mexico City Airport. For those not familiar with decision theory, this book should be read after an introductory text, such as Raiffa, *Decision Analysis.*

Maier, N. R. F., *Problem Solving and Creativity in Individuals and Groups*, Brooks/Cole, Belmont, California, 1970.

Surveys a number of research strategies designed to examine the efficiency of particular techniques in problem-solving groups.

Raiffa, H., *Decision Analysis: Introductory Lectures on Choices Under Uncertainty*, Addison-Wesley, Reading, Massachusetts, 1968.

An excellent introductory text on decision analysis. Decision theorists have focused on problem-solving stages and extending problem-solving paradigms from the individual to the group level.

Schein, E. H., *Process Consultation: Its Role in Organization Development*, Addison-Wesley, Reading, Massachusetts, 1969.

A book written for consultants called in to help small groups that are not functioning effectively. Chapters 4 and 5 are of particular relevance to members of problem-solving groups.

Zander, A., *Groups at Work*, Jossey-Bass, San Francisco, 1977.

A work by a leading social psychologist on some of the practical aspects of working in small groups.

6

Public Involvement as Planning Communication

Katharine P. Warner

Planners have acquired a reputation for developing plans that are not followed. Some of the reasons for this reputation have to do with planners' limited political authority and power. Other reasons are the difficulties many planners have in communicating their ideas persuasively to political decision makers, other agency administrators, and interested publics. Communication skills are essential for effectively involving members of the public in planning activities. Such skills include asking appropriate questions, listening well, and presenting information clearly and coherently.

To design an effective public involvement program that is an integral part of a larger planning activity, planners should initially address three major program elements: the objectives and role for public involvement and its related activities; the publics who might and should be involved; and the types of public involvement mechanisms that could be most effective in stimulating needed involvement.

WHY INVOLVE THE PUBLIC?

Planners collect, analyze, and present information about a wide variety of topics relating to land use, social and economic development, and environmental quality. A deliberate and carefully structured effort to involve publics in such activities is important to ensure that their concerns, preferences, and priorities are considered and discussed.

Urban and regional planners supported by public funds will generally find that public involvement in some form is legislatively and/or administratively required in their studies. As government programs have expanded in scope and size of resource commitments, the number of organized groups and general publics who see their interests affected and hence demand a voice in influencing planning recommendations have correspondingly multiplied. Two prerequisites for an effective participation process that have received particular attention from government officials and private and public interest groups are increasing public access to the types of information used in planning studies and including appropriate and procedurally guaranteed opportunities for the expression of public views and preferences.

Government planning agencies at all levels have been developing procedures and structural arrangements intended to satisfy the ever-expanding legislative and administrative directives regarding public involvement. The federal government has been a catalyst in this regard, using its funding and regulatory powers as leverage in the promotion of guidelines mandating citizen roles in plan formulation and review.

The Model Cities Program during its eight year existence proved to be one of the most far-reaching and complex organizational attempts to set up an effective public-planner partnership for formulating and implementing plans for a specific urban area. The program encountered numerous problems but also developed a number of innovative mechanisms and precedents for participatory planning. The U.S. Department of Housing and Urban Development issued a set of performance standards for citizen participation in Model Cities programs that stressed such factors as creating an organizational structure whose leadership and activities would be accepted as representative by neighborhood residents and providing neighborhood groups with sufficient information to enable them to initiate proposals and respond knowledgeably to those of others. Technical assistance was to be provided in a manner agreed to by the neighborhood groups. Finally, it was recognized that financial hardship might prohibit poor people from becoming involved in planning efforts so agencies were permitted to compensate people for participation (U.S. Department of Housing and Urban Development, 1967). Model Cities agencies used a variety of approaches in responding to HUD's performance guidelines. These included employment and training of neighborhood residents as planning aides and community fieldworkers and development of subcontracts with neighborhood groups to enable them to hire their own consultants.

The Housing and Community Development Act, which authorized the Community Development Block Grant (CDBG) program, also administered by the U.S. Department of Housing and Urban Development, explicitly calls for giving citizens adequate information about the program, holding public hearings to obtain citizens' views, and providing citizens with an adequate opportunity to participate in developing a locality's program application. In subsequent performance guidelines for the CDBG program, HUD specified that a citizen participation plan should be developed and made public by the local community development administrative agency. This plan was to include a timetable showing when and how information would be disseminated and public hearings held, when and how any technical assistance would be provided, and when and how citizens would participate in plan revisions. In addition, localities were required to document and maintain records on their public participation activities.

Other federal agencies influencing or directly engaged in planning activities, such as the U.S. Department of Transportation, the Environmental Protection Agency, and the Army Corps of Engineers, have issued an extensive array of directives and guidelines on public participation. Recent federal-planning-related legislation containing specific public involvement requirements includes the Airport and Airways Development Act, the Federal Water Pollution Control Act, the Coastal Zone Management Act, and the Energy Reorganization Act.

Public involvement legislative initiatives are not limited to the federal level. For example, in a survey of 230 state, regional, and local public planning agencies of various types located throughout the country, over 80 percent of those responding indicated their agency did operate under specific legislative or administrative regulations requiring some type of public involvement, review, or consultation procedures (Warner, 1971). That proportion undoubtedly continues to increase. A 1975 study of Pennsylvania government showed that 165 state statutes contained requirements for advisory committees and public hearings. The study also estimated a yearly expenditure of between $50 million and $100 million by state agencies to carry out citizen participation activities (Stuart Langton & Assocs., 1976).

OBJECTIVES FOR PUBLIC INVOLVEMENT

The communication of facts, ideas, and opinions among planners and publics can build a mutual awareness of problems and needs, which in turn serves as the basis for the development of politically acceptable solutions. Thus public involvement can be an important source of information, both factual and attitudinal, for planners. The process should be structured to provide publics with the background information necessary to form judgments and to express preferences where appropriate. Finally, opportunities for the collaborative exchange and discussion of information among planners and publics are important both in refining planning data and recommendations and in developing support among affected publics for study efforts. There are three basic objectives for structured public involvement:

1. Expanding the amount and usefulness of information available to both planners and publics. This would include providing, receiving, and exchanging information.
2. Providing a fuller opportunity for publics to affect and influence planning recommendations. This would include evaluation sessions, as well as opportunities for formal or informal negotiations among various participants in the process.
3. Development of public support for planning recommendations. Such support depends on the perceived credibility of the planners and the projected effects of the project.

The achievement of these objectives, shown in Figure 6.1, requires a substantial commitment of planning resources and communication skills.

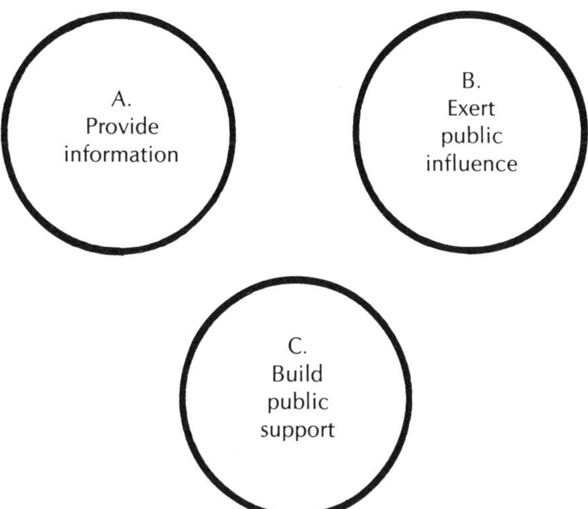

Figure 6.1 Objectives for public involvement.

The public hearing on City Opportune's new transportation plan was crowded and vitriolic. Among the many complaints planners listened to were that the plan had very little input from the public; was overly biased toward building more highways; was insensitive to environmental considerations (these included everything from air pollution to parks and open space); deliberately targeted low-income, powerless groups for route dislocations and did not adequately address public transportation needs of the elderly, poor, and handicapped; lacked consideration of some transportation alternatives, including bicycles; relied too heavily on the continued availability of large amounts of federal funds; and used biased information and misinterpreted data to justify project recommendations.

The next day, the Opportune planning staff met to decide on a future course of action. During the meeting, it was noted that the Citizens' Association for Neighborhoods intended to file suit enjoining the city from submitting the present plan in applications for federal funding because, they maintained, it had not complied with the environmental impact statement requirements for consideration of alternatives. Based on the testimony of the preceding evening, the planners believed they could easily lose the court case, so they decided to revise the present plan using a more comprehensive approach. This time they would seek public involvement, starting with the transcripts from the public hearing.

Because of the numerous groups expressing interest in transportation, the planning staff decided to establish a system to ensure that information flow to and from these publics was encouraged, a sufficient variety of opportunities for involvement was provided, and the full range of different viewpoints was represented. In this way, the staff hoped to inform themselves better about the needs of constituents and in turn to make their publics more aware of the financial constraints looming ahead. The planners realized that their final recommendations would have to strike a better balance between needs and available resources if they were to be politically supportable.

DEFINING THE PUBLIC

The word *public* is often used as if there were a single entity, that is, one group or category of individuals who share a common and distinctive interest and set of preferences. In fact, there are myriads of distinctive publics who could be involved in any major planning study. Most do not become active for a variety of reasons—for example, lack of knowledge, interest, resources, or opportunities or because they trust the capability of others involved, either professionals or fellow citizens. In this chapter, the term *publics* will be used to designate a broad range of groups and individuals whose perspectives, interests, and/or responsibilities differ from the immediately involved planning unit.

Political Officials and Government Staff Members

Public-sector representatives often are not included among those traditionally categorized as publics. However, their participation or interest should never be taken for granted but rather carefully provided for in the design of a public involvement program.

Elected officials tend to be busy and oriented more toward immediate crises than longer-range planning activities. For obvious reasons, they are quite interested in the opinions of their constituents and the intensity with which these are held. An effective public involvement program is, in some respects, doing the politicians' canvassing work for them. For this reason and because politicians are often contacted by members of the public to register complaints or to find out what is going on, planners should be careful to keep such political officials well informed about their planning activities. Certain politicians, because of special experience, expertise, or interest may even want to participate in developing the plan.

Because government agencies, particularly at the local and state levels, are likely to be primary consumers and users of planning information, it is essential to include them in planning studies. This will enable such organizations to understand what information is available, how they can obtain it, and how it might be applicable to the work they are doing. For example, information on land cover and uses collected by a planning staff might be of immense value to the public works department, the drain commission, the road commission, the parks and recreation department, and the county extension and soil conservation offices.

In turn, these same user organizations may be rich sources of helpful information. In many cases, they are also likely to play important roles in the eventual implementation of adopted plans. For all these reasons, it makes sense to involve other government agencies early and throughout the formulation of plans.

Private Participant Publics

Publics can most easily be identified in relation to particular issues. Whether they will become involved in the consideration of these issues depends on at least three factors. The first is the degree to which the individuals and the organizational leaders making up these publics are aware of the issues to be decided in the planning process and whether they perceive their interests being affected by the types of decisions made. The second is the degree to which people believe they can influence the formulation of the plan on their own behalf. And finally, because each person's and organization's resources are limited, the costs of participation in time and energy will be evaluated against the benefits that could potentially result from their involvement, or conversely, the benefits that might be lost if they chose not to participate.

Surveys of public opinion and voter attitudes have shown that people's awareness of issues and their implications are related to their socio-economic status. Those with higher levels of income and education consistently indicate greater knowledge of potential events and their likely consequences. In addition, such people also show a greater confidence in their own ability to influence the system, and their levels of political and community participation are correspondingly much higher (Milbrath, 1965; Almond and Verba, 1963).

Membership in organized community groups is also related to higher rates of community participation. Such groups provide an opportunity for individuals to communicate more frequently with each other regarding issues, thus raising their levels of interest and knowledge. At the same time the organizations provide a mechanism for taking action and exerting influence on governmental decision makers.

Typically participants in planning studies tend to be drawn from the following types of private groupings: private organizations with an economic stake in the potential planning results; private groups organized around a particular common interest, such as recreation or community development; public-interest groups such as the League of Women Voters; those who have developed a particular expertise or interest in affected planning issues, such as technical experts and/or community opinion leaders; and persons who may be directly affected by project developments, such as landowners. Often such participants can provide planners with new information or suggest additional management solutions based on their own experience, firsthand knowledge of a geographic area, or technical expertise in a particular subject area. Because public-interest groups are often viewed as capable, objective, and nonpartisan, they can be valuable sponsors of or provide chairpersons for public involvement activities.

Methods of Identifying Publics

For planners seeking to involve a representative and useful range of private groups and individuals in their efforts, it is important to identify potential publics at the beginning. This is neither simple nor easy, and planners should be alert to potential biases. Knowing who probable participants are likely to be is the first step toward developing effective mechanisms for acquiring and disseminating information. Such an awareness also enables planners more consciously to seek the participation of groups that tend to be underrepresented or disadvantaged in processes involving the analysis of technical information.

Planners should seek to publicize opportunities for participation as widely as possible so that everyone with a potential interest in becoming involved has the chance to do so. Mass media probably offer the best way to accomplish this. It is also important for planners to identify those publics (organizations and specific individuals) who are especially important to include as sources of information and feedback and as representatives of various interest groups or organizations, such as environmentalists, affected industries, landowners, and clients.

Information about potential participants can come from three types of sources. The first is through self-identification in which interested persons are encouraged to make themselves known directly through letters, telephone calls, visits, testimony, and other avenues. Those opposed to certain actions may announce themselves through petitions, protest demonstrations, or even lawsuits.

Second, the planning staff may seek to compile a listing of appropriate population groupings, organizations, and individuals. Sources for such staff identification include organizational membership and officer lists, social and economic profile statistics for an area (which could form the basis for special surveys or publicity efforts), newspaper files, and city directories.

Finally staff members may ask particularly well-informed persons, organizational officers, and others for participant suggestions. Such third-party references may considerably expand the potential number of participants. But to guard against professional biases, it is important to consult a diverse set of reference people.

Publics as Opinion Leaders

Public opinion research has revealed that certain members of the public assume the role of key communication linkages with regard to information disseminated by the mass media and through official government and private organizational channels. This process has been called the two-step flow of communication, and the linkage people have been termed opinion leaders (Katz and Lazarsfeld, 1955). These people transmit the information disseminated by the media and official sources directly to those with whom they interact, face to face. Such people in turn interact with others, setting off a communication chain-reaction process.

As the information is transmitted, it is interpreted by the opinion leaders and those with whom they interact. Thus opinion leaders play a key role in shaping public reaction to various programs and proposals described by technical experts or the mass media. For example, an opinion leader's interpretation of original information can affect the degree to which additional coverage is listened to and believed.

Opinion leaders may be defined using a combination of three approaches: positional, decisional, and reputational. A positional list of opinion leaders is compiled by recording people who hold formal leadership positions in various private and public organizations. A second decisional list would consist of those who have been active and taken advocacy positions on important planning issues. Area newspaper files are usually the best source for such names. Finally, those interviewed from the first two lists are asked whose opinions would carry great weight locally if a decision had to be made on a planning issue of the type being studied. Those nominated by at least three persons are then listed as reputational candidates for involvement. As the lists expand, they will tend to overlap (Borton et al., 1970).

Opportune's planning staff studied the public hearing transcript in preparing a new work program for reviewing and revising their draft transportation plan. Since more effective public involvement was a primary goal, they sought to identify various individuals and groups with whom they should exchange information early in the revision process. Among those groups actively involved at the public hearing were advocates for elderly and handicapped persons, several environmental organizations, a representative of the Transit Workers Union, the president of the Downtown Business Association, the president of the League of Women Voters, a school PTA officer, and several persons from the Citizens' Association for Neighborhoods and other neighborhood organizations.

The staff compiled a potential participant list starting with the names of people who registered at the public hearing and adding to these the names of all city, county, state, and federal politicians with responsibilities in the area under discussion. For the original study, the staff had put together a loose coordinating committee of people in local government agencies who had fairly direct transportation-related duties; these included the director of the local transit authority, the county engineer, and others. Staff now enlarged this listing to include those with governmental responsibilities that might be affected by plan implementation and/or those who might provide additional specialized information that could affect project feasibility; these included the county drain commissioner and the city, county, and regional parks directors. Government agencies at the state and federal levels, as well as those representing transportation-dependent user groups such as the Area Agency on Aging, were also added.

Finally the staff expanded the list to include all major civic, neighborhood, and special interest groups likely to have some interest in the revised transportation plan. Whether the groups would choose to participate was left up to them, but the planning staff was committed to providing them with that choice. Whenever possible during their initial public contacts, the staff inquired about other potential participants who might be included in planning activities.

HOW TO INVOLVE THE PUBLIC

Communication Purposes for Involving Publics

Public involvement activities can serve three major communication purposes for the planner: the one-way presentation of information to publics; the one-way receipt of information from publics, often termed feedback or simply reaction, and exchanges of ideas and opinions that build upon shared initial information as the ideas evolve.

Public involvement mechanisms typically reflect one of these communication functions most strongly. But they can also be combined as part of a program sequence to complement and support each other. For example, people need access to information about a study and its objectives before they can provide meaningful reactions to planners' proposals. Participants, in turn, may integrate their own experience and preferences with the study's information to suggest new ideas for planners to consider. Figure 6.2 shows a number of commonly used public involvement mechanisms classified by their primary communications purpose.

Features of Mechanisms to Involve the Public

Mechanisms that present information

Public involvement mechanisms that emphasize dissemination of information from planners to publics include use of the mass media (newspapers, television, radio, and films), written reports, newsletters, and planner presentations and speeches. Staff persons working with these types of informational presentations have frequently been labeled the public relations group. Often they have worked previously in the media, and sometimes their products are somewhat derisively referred to as publicity. Today this stereotype is changing as people increasingly recognize that, if publics are to become contributing members of the planning process, they must have a foundation of planning information on which to base their activity. For example, the New York Regional Plan Association

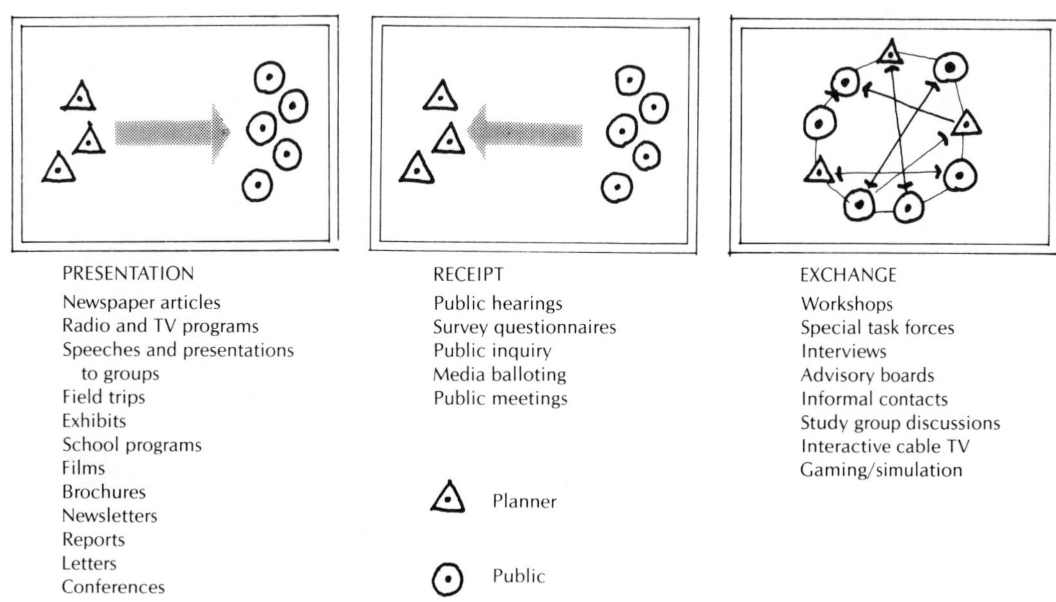

PRESENTATION
Newspaper articles
Radio and TV programs
Speeches and presentations to groups
Field trips
Exhibits
School programs
Films
Brochures
Newsletters
Reports
Letters
Conferences
RECEIPT
Public hearings
Survey questionnaires
Public inquiry
Media balloting
Public meetings
EXCHANGE
Workshops
Special task forces
Interviews
Advisory boards
Informal contacts
Study group discussions
Interactive cable TV
Gaming/simulation
Planner
Public

Figure 6.2 Primary communications functions of public involvement mechanisms.

has developed a set of background booklets and half-hour television programs to precede and inform an extensive program of community study group discussions regarding futures for the region.

The mass media are well suited to establishing a basic awareness of problems and issues and how they are being studied. By themselves, the media do not provide a sufficient basis for public involvement; however, it is crucially important for the planner to understand how to use media well and creatively as a first step in the process.

The media, especially newspapers and television, are deluged with press releases and requests for coverage. Planners who want to attract their interest and to encourage accurate coverage should meet with media representatives and personally try to interest them in the topic. Media persons should be recruited like any other group of active public participants.

Since most journalists and media production people are not specialists in planning matters, it is important for planners to provide them with a personal background briefing, as well as with appropriate high-quality content materials, which can include succinct and publishable news releases for newspapers, and high-quality visuals, such as glossy photos, maps, models, and film coverage for television stations.

Planners who are presenting planning information to the media should consider what will be of interest from their perspective. It is also essential for them to assess the type and size of the audience they are likely to reach so they can prepare appropriate follow-up activities ahead of time. Planners should be sure media people have a readily accessible contact person (name and telephone number) to answer further questions and/or provide a review of the story.

Speeches and presentations at organizational meetings can often be an efficient and effective use of limited time and staff resources. The availability of staff to carry out these efforts must be made known to potentially interested groups. Planners, when invited, are likely to find an interested audience from which future participants may be recruited.

Written publications are a common and highly structured means of communicating planning information. In developing a useful set of materials, the diverse technical backgrounds, levels of interest, and education of potential reader publics should be carefully considered.

Mechanisms for receiving information

Mechanisms especially oriented toward getting informational feedback and reactions from publics form the core of what has traditionally been considered to be public involvement. These include public hearings, survey questionnaires, and certain types of public meetings. The feedback function they perform for planners is important. However, unless they are part of a sequence and combination of related public involvement activities, their informational value is often greatly diminished because the publics responding may lack a frame of reference and relevant background information on which to base their answers. Also the responses are highly individualistic and tend to be unrelated to one another in terms of ideas and suggestions.

Public hearings are both the most common and most maligned public involvement mechanism used by planners. They involve one-way communication (publics to planners), which is often adversarial. Because of their usual format of individual speakers presented in order of status (such as elected officials, appointed officials, representatives of organized groups, and individuals) or in order of request for time, there is no iterative development of issues or ideas, and such hearings tend not to be a good means of refining planning information. The written testimony, however, is often well researched and thoughtfully put together. A major drawback in its use is that typically public hearings occur at the very beginning and end of plan development, both times ill suited for planners to make use of specific testimony.

Although public hearings suffer from deficiencies as information sources, they fulfill an important function for public involvement: they serve to meet the guarantee of due process and offer a publicized opportunity for all citizens to state an opinion that is documented as a matter of public record.

Attitude and opinion surveys are among the best ways to contact a broad range of people, thus cross-checking the information secured from those publics that participate most actively in organized events. Such surveys need to be carefully designed to be meaningful to the respondents and provide planners and designers with useful information. They also tend to be expensive to administer and analyze, so the appropriateness of their use should be carefully considered.

Mechanisms for sharing and exchanging information

Public involvement mechanisms that provide for two-way communication between planners and publics are potentially the richest source of useful planning information. Such mechanisms include workshops, charrettes, task forces, and citizen advisory boards. Their objective is to provide reasonably informal and open meetings at which professionals and interested publics can exchange information. In these settings, people are able to refine information, discuss issues, and ask questions about planning topics of key importance to them. The special knowledge or expertise of various participants can thus be melded or cumulatively built upon.

Three cautions are pertinent when discussing interactive mechanisms. First, they are most effective when small working groups are used, thus limiting the number of people actively involved. Second, they are highly staff-intensive. That is, to be effective, interactive mechanisms require thorough staff planning as well as preparation of supporting materials. In addition, staff participation as the professional half of the desired interaction requires both considerable time and skill in managing group dynamics. Finally, interactive mechanisms require substantial commitments of time and interest on the part of participant publics. This both automatically limits the number of people likely to get involved and introduces the possibility of considerable bias in the makeup of active publics. Thus those with the most resources and the greatest potential benefit at stake will tend to be overrepresented.

Citizen advisory boards are the most often used interactive public involvement mechanism. They offer considerable potential benefits both as a source of differing perspectives and as an auxiliary planning and community relations unit. They also pose some difficult questions about the public involvement process. How should members of such advisory bodies be selected? In what sense can the members be considered representative? How can they be made accountable to a larger public constituency? What roles and responsibilities should be assigned to such groups in plan formulation?

Members of an advisory committee may be selected in a number of ways. The most common methods include appointment by the chief elected official, such as a mayor, and designation as the representative of a particular community organization by its own membership. Regardless of the method, at the start of its activities, planners should be careful to define explicitly the roles, responsibilities, and prerogatives of the committee. Members should clearly understand whether they are to make decisions or offer advice, and in the latter case to whom.

Providing citizen committees with needed staff support can consume large amounts of time and resources; however, it is a necessary requirement to produce useful information. Some planning agencies have actually given their citizen advisory committees managerial and oversight responsibilities for public involvement programs, feeling they would be able to gauge information and activity needs from a more informed and objective standpoint.

Informal contacts are a form of interactive communication frequently forgotten when establishing formal, structured, public involvement programs and processes. They include instances where planners informally seek advice, reactions, or ideas from others or where members of the public seek out planners to raise questions or present ideas.

Informal contacts tend to be highly individualistic and entirely discretionary. Thus they present certain problems from a public involvement standpoint. One is that they are frequently invisible. That is, a planner may be consulting frequently with various local officials and civic group leaders, but such activities are not readily apparent to other publics. Also, informal contacts provide no publicly well-defined and structured means for securing citizen input; for example, an agency may depend on only those citizens who complain or ask questions. The lack of structure and the broad administrative discretion inherent in the use of informal contacts may discourage certain publics from participating because they are unaware of such opportunities or they believe planners are interested in listening only to certain groups.

Workshops are useful for providing face-to-face small group discussions between planners and publics, especially community opinion leaders. A number of formats may be used. One is an initial plenary session introducing the status of work to date, followed by division of participants into smaller working groups, usually ten to fifteen people, which report back to a concluding plenary session.

The major work occurs during the small group discussions, which may be organized around a particular problem topic, such as commercial redevelopment, or a specific geographic locale, such as the West Side neighborhood, or left with an open agenda for participants to determine. Planners should take part in these discussion groups as technical resources and participants. However, it is often better to have a member of the public assume the role of chairperson. Maps, appropriate data summaries, and sketch paper taped on the wall are useful working tools as the group considers various planning alternatives.

Interactive cable television does not yet appear to have realized its full potential for public involvement. However, both the 1972 ruling by the Federal Communications Commission requiring that cable systems installed in the one hundred largest television markets must provide a channel for municipal services and an eventual capacity for two-way services, and the explosive growth of U.S. cable coverage have enhanced the possible usefulness of this communications technology for planning.

Various formats to involve the public could be developed by planners, including the polling of viewers by digital signal following a live or videotaped presentation of planning issues, holding of teleconferences where groups of officials and citizens at various locations compare ideas and discuss their views, and feedback of immediate public response to televised planning commission or city council meetings from various community viewing locations such as public schools, libraries, and senior centers.

With a growing list of potential public participants in hand, the Opportune planning staff turned to the problem of how these publics could most effectively be involved in the plan revision process. The planners saw an immediate need to disseminate information describing the revised planning study's scope and schedule. Announcements by the mass media and an article in the local newspaper describing the study as well as a mailing to people on the "preliminary publics" list, were decided on. The staff also believed members of the public should be aware of and have access to the large quantity of background data already collected. For this purpose, they set up a depository collection of materials in the main public library and developed a series of exhibits for display in libraries, schools, community centers, and other public buildings.

The planners decided that in addition to presenting study information to the public, they needed to discuss program implications of this information with community opinion leaders and interest group representatives. The public hearing had demonstrated that the planners' interpretations differed substantially from those of many of these people, and a means for reconciling the differences and working out shared priorities was clearly needed. Various possible public involvement mechanisms were considered, such as workshops, study groups, task forces, charrettes, informal contacts, interviews, and even establishment of a citizen advisory committee. The pros and cons of each were discussed, but a final selection of program mechanisms was postponed pending a thorough analysis of the staffing and financial feasibilities of various combinations.

At the public hearing, the draft plan was repeatedly criticized for underplaying the role for public transit service. The planners admitted this was true, albeit reluctantly, because financial resources to support transit seemed limited. However, the staff decided to survey the local public to see if comments at the public hearing had accurately reflected local sentiment and a majority of people might be willing to approve a local transit tax.

Operational Characteristics of Mechanisms to Involve the Public

Designing a public involvement program that complements technical planning work tasks requires an appropriate choice of communication mechanisms. This choice depends upon study objectives, the scope of work, the resources (staff time and money) available for the job, and the complexity of the problems being studied. For example, transportation planning for a heavily urbanized area with many jurisdictions and several transportation modes would involve more diverse publics and a more extensive media dissemination effort than planning for a single neighborhood's street improvements. Once a planner has identified the information needs and communication objectives for a planning study, the classification scheme in Figure 6.3 can be useful in choosing

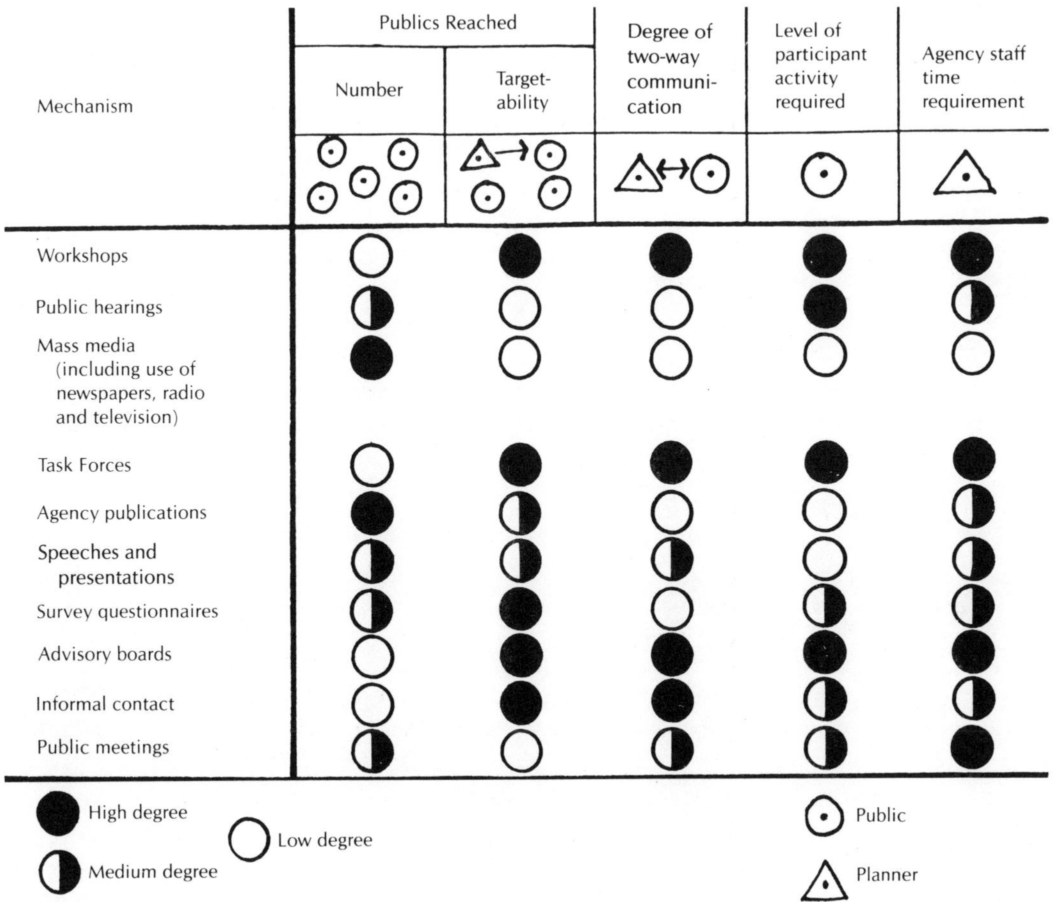

Mechanism	Publics Reached: Number	Publics Reached: Target-ability	Degree of two-way communi-cation	Level of participant activity required	Agency staff time requirement
Workshops	○	●	●	●	●
Public hearings	◐	○	○	●	◐
Mass media (including use of newspapers, radio and television)	●	○	○	○	○
Task Forces	○	●	●	●	●
Agency publications	●	◐	○	○	◐
Speeches and presentations	◐	◐	◐	○	◐
Survey questionnaires	◐	●	○	◐	◐
Advisory boards	○	●	●	●	●
Informal contact	○	●	●	◐	◐
Public meetings	◐	○	◐	◐	●

Figure 6.3 Operational characteristics of public involvement.

appropriate public involvement mechanisms. Knowing the strengths and requirements of each enables planners to use them more effectively.

The characteristic of "publics reached" has two dimensions: the number of people who can be contacted and the agency's ability to communicate with particular people or groups, known as *targetability.* Two-way communication between planners and publics usually produces more issue-related or new environmental information for planning studies. One-way mechanisms are most useful in either distributing study information or receiving evaluative comments.

"Level of activity required from participants" refers to the amount of initiative and commitments of time and energy required from publics for the mechanisms to be successful. Before a final set of program mechanisms is agreed to, it is important for the planner to assess the characteristics of the publics in the study area.

The agency staff time requirements are a major constraint on the size and composition of any public involvement program. Sometimes the resources needed for certain mechanisms to work well are underestimated, leading to frustration with the program operation and results.

Public involvement is not a one-time, one-event occurrence but rather a process for introducing and considering the ideas of those affected by planning throughout the time a plan is being developed. The process needs to include combinations of different mechanisms that reinforce and complement each other. For example, individuals should have already received basic information about a study, its purpose, and the initial findings to participate effectively in a planning task force meeting or workshop. Various information dissemination mechanisms like brochures and reports, newspaper articles, and study newsletters can be of considerable value in providing such background. Following the discussion meeting, a brief mailed summary of major ideas brought out and the action that the planning agency expects to take in regard to them can prove useful in reinforcing participants' interest in the study and their willingness to participate in further activities. This example of reinforcing public involvement activities could contain the elements shown in Figure 6.4.

Having established their informational needs and laid out a comprehensive list of potential public involvement activities, the planners began the difficult task of selecting a set of activities that could be performed well, given their present resources, and would be expected to yield the information and public support needed.

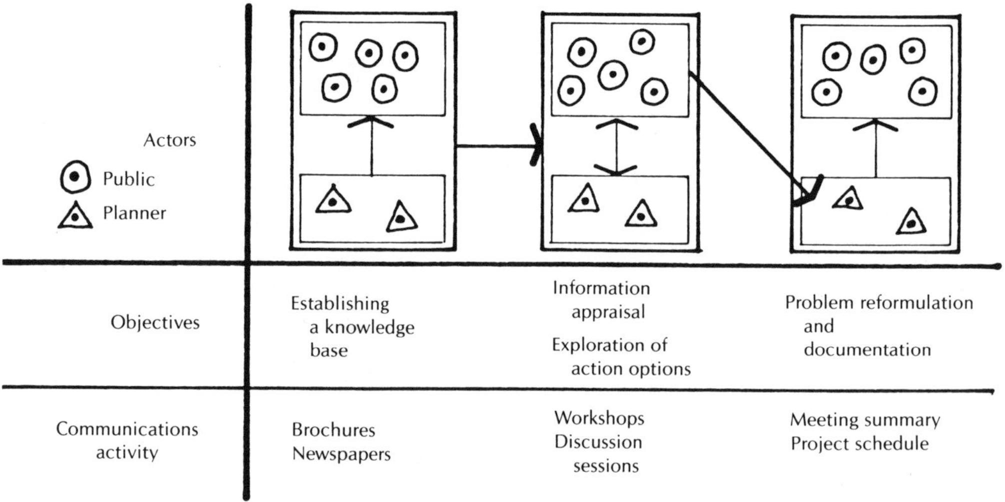

Figure 6.4 Communications elements.

An initial community opinion leader workshop, which the League of Women Voters agreed to sponsor and help arrange, was scheduled. A large number of informal contacts were to be made, and an initial community public meeting was held with both a panel of speakers and considerable time alloted for public comments. These activities were expected to help the planners identify key issues the plan would have to address to warrant public acceptance. In addition, new information on transportation problems and differing interpretations of the data were expected to be revealed.

A survey designed to validate certain issues raised at both the original public hearing and the first round of new involvement activities was first thought to exceed the planning agency's available personnel and budget. However, a survey research class at Opportune College agreed to help design the survey and carry out the fieldwork at minimal cost, so it became possible to include it in the revised program.

The planners believed public discussion and information would be especially critical at the time in the study when alternative programs and management options were being developed. Criticism of the draft plan had centered on inadequate consideration of differing approaches. In response, the planners programmed a series of workshops, informal contacts, and a mail-in questionnaire that would accompany a second newspaper story discussing background issues. To ensure that the workshops would be effective in generating ideas, the planners built into their budget the time and staff costs necessary to produce premeeting informational materials, postmeeting summaries, and follow-up evaluations.

After a recommended set of transportation proposals had been drawn up, the planners scheduled several public meetings similar to the initial one in format. In addition, they built into the program both written materials describing the proposals and background mass media coverage. It was expected that certain revisions and refinements in the proposals would be made following this round of evaluative public involvement.

The planners emphasized the importance of continuing media coverage as well as staff contacts and public access to information throughout the study process. Thus frequent presentations and speeches to interested community groups were projected, and one staff member was specifically assigned the job of liaison with the media and respondent to public inquiries, suggestions, and complaints. (See Figure 6.5.)

SUMMARY

An effective participatory planning process should enable planners to test the social acceptability of their assumptions and program proposals, as well as enabling involved members of the public to express their preferences among potential management alternatives. To date, there have been few attempts by planning agencies to define performance criteria for evaluating the effectiveness of public involvement activities. Such a list might start with the following general criteria:

1. The planning process should provide opportunities for members of the public who choose to participate to do so.
2. The publics should be made aware of the availability of such participation opportunities so they can make a choice.
3. Sufficient information should be made available to the publics so they can participate effectively.
4. Planning agencies should be able to respond effectively to the inputs and activities of public participants.

There are no hard-and-fast best procedures for structuring public involvement. Those arrangements that will work most effectively depend to a great extent on the specifics of each situation, such as the focus and geographic scope of the planning effort, the resources available to planners, the history and seriousness of planning problems in the area, the types of area civic and special interest groups that are active, and the degree of importance they attribute to the planning efforts.

Building an effective public involvement process is more an art than a science. Planners need to develop the best program communication mosaic for each situation. Care should be exercised to maintain program balance and an openness to different opinions so as not to screen out certain interest groups. Ignoring either the noisy or silent opposition only postpones the conflict and often increases its later intensity.

To build an open and effective framework for public involvement, the planner needs a variety of communication skills, including the abilities:

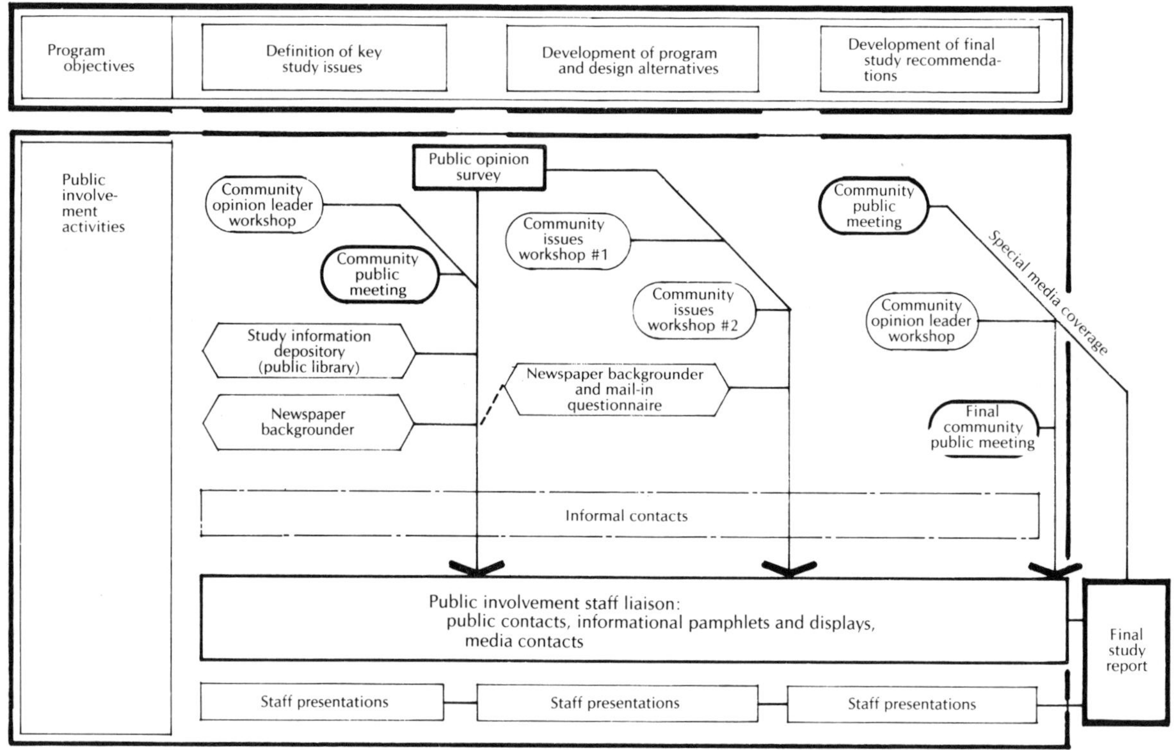

Figure 6.5 Opportune's program design for public involvement.

1. To write reports, articles, brochures, and press releases clearly.
2. To use graphics of all types to illustrate the kinds of information being presented either verbally or in writing.
3. To present verbally the essential facts behind planning recommendations.
4. To frame alternatives or planning options in a manner that enables people to respond meaningfully and state their preferences.
5. To formulate, carry out, and analyze surveys so that they provide useful planning information.
6. To listen to people and answer questions in a manner that is both sensitive and responsive.
7. To participate productively in group discussions.
8. To recognize and understand the social and political implications of various planning contexts.

APPLICATIONS: DESIGNING A PROGRAM FOR PUBLIC INVOLVEMENT

1. Assume the role of urban planner for a community you are familiar with. The mayor has asked for a revision of the fifteen-year-old land use plan of the community. As the junior planner on the staff, you have been asked to develop a proposed public involvement process to accompany other plan revision activities. How would you do this? What would be the principal elements of this process? Incorporate specific objectives, publics, mechanisms, and timing.
2. Your department director, who is skeptical about the benefits of public involvement, has asked you to address the following three questions:
   a. What types of information from local res-

pondents are needed to increase the plan's effectiveness?

b. What kinds of usable information are local respondents able to provide, given their probable levels of technical knowledge about planning possibilities, opportunities, and implications?

c. How should the information elicited through your public involvement program be analyzed and incorporated in the planning process?

REFERENCES

Almond, Gabriel A., and Sidney Verba, 1963, *The Civic Culture,* Princeton University Press, Princeton, New Jersey.

Borton, T. E., K. P. Warner, and J. W. Wenrich, 1970. *The Susquehanna Communication-Participation Study,* U.S. Army Corps of Engineers, Institute for Water Resources, Washington, D.C., NTIS, Accession No. AD717023.

Katz, Elihu, and Paul F. Lazarsfeld, 1955, *Personal Influence: The Part Played by People in the Flow of Mass Communication,* The Free Press, Glencoe, Illinois.

Langton, Stuart, and Associates, 1976, *A Survey of Citizen Participation Requirements and Activities Among Major State Agencies of the Commonwealth of Pennsylvania,* Fremont, New Hampshire.

Milbrath, Lester W., 1965, *Political Participation: How and Why Do People Get Involved in Politics?* Rand McNally, Chicago.

U.S. Department of Housing and Urban Development, 1967, Demonstration City Letter No. 3, October 30.

Warner, K. P., 1971, *A State of the Arts Study of Public Participation in the Water Resources Planning Process,* U.S. National Water Commission, Washington, D.C., NTIS.

BIBLIOGRAPHY

Arnstein, Sherry R., "A Ladder of Citizen Participation," *Journal of the American Institute of Planners* **35**:216-224 (July 1969).

Provides a typology of citizen participation based on the degree of influence exercised by public participants.

Borton, T. E., K. P. Warner, and J. W. Wenrich, *The Susquehanna Communication Participation Study,* U.S. Army Corps of Engineers, Institute for Water Resources, Washington, D.C., 1970, NTIS, Accession No. AD717023.

Describes a major public involvement evaluation study done in conjunction with river basin planning. Contains a manual for developing public information workshops.

Burke, Edmund, "Citizen Participation Strategies," *Journal of the American Institute of Planners* **34**:287-294 (September 1968).

Provides a typology of public involvement based on the planning agency's objectives.

Godshalk, David R., and Bruce Stiftel, "Making Waves: Public Participation in State Water Planning," *The Journal of Applied Behavioral Science* **17**:597-614 (No. 4, 1981).

Discusses public participation in the Federal 208 water resource planning program in terms of policy lessons. Proposes an exchange model for evaluating participation programs.

Hyman, Herbert H., "Planning with Citizens: Two Styles," *Journal of the American Institute of Planners* **35**:105-112 (March 1969).

Recounts and contrasts two planner approaches to citizen participation in connection with urban renewal projects.

Langton, Stuart, ed., *Citizen Participation in America: Essays on the State of the Art,* Lexington Books, Lexington, Mass., 1978.

State-of-the-art discussions of important citizen-participation program types, mechanisms, and issues.

Marshall, Patricia, ed., *Citizen Participation Certification for Community Development: A Reader on the Citizen Participation Process,* National Association of Housing and Redevelopment Officials, Washington, D.C., 1977.

Emphasizes citizen participation aspects of the Community Development Block Grant program. Contains many articles describing the use of different public involvement mechanisms.

U.S. Federal Regional Counsels and Community Services Administration, *Citizen Participation,* Government Printing Office, Washington, D.C., January 1978.

Identifies and discusses federal requirements and regulations for citizen participation.

Warner, K. P., "Communication of Environmental Information," in W. Marsh, ed., *Environmental Analysis for Land Use and Site Planning,* McGraw-Hill, New York, 1978.

Discusses key aspects of the public involvement communications process: audiences, formats, and mechanisms.

Warner, K. P., *Public Involvement in Water Resources Planning,* U.S. National Water Commission, Washington, D.C., 1971, NTIS, Accession No. PB204245.

A review that discusses publics, program guidelines, and mechanisms. Includes case examples and surveys of public agencies and citizen groups.

7

Planning and the Computer

Mitchell Rycus

Computers allow planners the control of and access to large amounts of information. The computer, increasingly important in all phases of information processing and communication, has been available for about thirty years, but recent breakthroughs in prices, size, speed, capacity, and intercommunication among computers will soon make them available for every office.

In the Introduction to this book, Junior Planner soon learned that the downtown Main Street garbage pickup day was Tuesday. He needed that information on the weekend when the city's environmental engineering department was closed. So he went to his office, turned on something that looked like a small television with a typewriter attached to it, and waited until the message "LOGIN PLEASE" appeared. He then typed LOGIN PLANNING, and hit the key marked RETURN, just like on any electric typewriter. Soon some characters and words appeared, letting him know that the computer was on and waiting for his instructions. He then typed SLIST GARBPICKUP, pushed the RETURN key, and information began to appear on the screen line by line. He noted the following display on the screen:

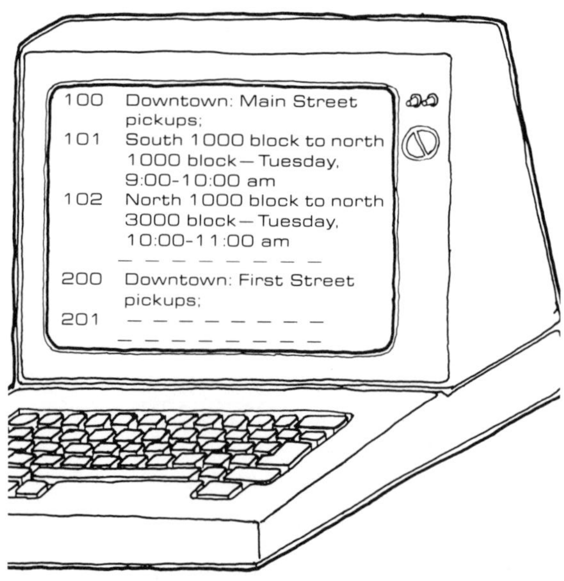

and he wrote on a pad 100, 200. When the lines stopped appearing on the screen, he typed SPOOL GARBPICKUP 100-200, and then, LOGOUT, turned the console off, and walked to another machine, which had begun to print out information. He waited until the printer stopped, tore off the sheets of paper, and, making sure he had a printed record of the schedule, put them into a manila folder marked "Main Street Garbage Analysis." The entire process took less time than making a telephone call during normal working hours to the city's environmental engineering department requesting the same information.

Later, after his initial analysis of all the data including the garbage schedule, Junior Planner went back to his computer console and typed:

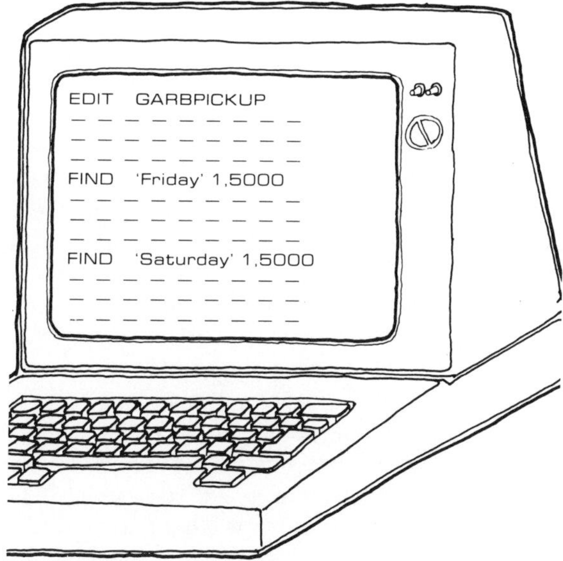

The computer printed all the lines in the file having Friday or Saturday garbage pickup information, and from this information Junior Planner was able to support his recommendation to change the Tuesday garbage pickup to Friday or Saturday.

This procedure may vary in detail from city to city and in the type of computer used, but in essence it would be very similar. Computers have greatly increased a practicing planner's access to secondary sources of information. A planner can now obtain more information more cheaply and quickly and have it displayed in a useful form. Junior Planner simply requested some information he knew was stored in the computer's memory and got it. His requests were typed in sort of a shorthand English. SLIST means that the machine is to list a file; in this case, the file's name was GARBPICKUP. The computer responded by listing every line in the file called GARBPICKUP. Later the planner was able to take advantage of the computer's analytic capabilities by requesting it to let him examine (EDIT GARBPICKUP) the file in greater detail. Instructing the computer to FIND all the lines where the word FRIDAY, or SATURDAY occurred, from the first line to line number 5000, is a routine procedure. It was also easy to tell the computer to print out (SPOOL) only lines 100 through 200 of the file on a printing device controlled by the computer. As a result, only that information which the planner needed could be selected and obtained by him without any need to involve other people.

COMPUTER LANGUAGES

Computers interpret combinations of words, numbers, and characters by converting them to some numerical value. Then depending upon the value computed, logic circuits are activated to process the information.

One can use two levels of communication for operating the computer. One level is designed for a specific computer's operating system and is referred to as *internal commands*. The other level is through a program written in an interpretable computer language that causes some specific set of operations to be performed.

Junior Planner used internal commands interactively to get the computer operating. The first command he entered was LOGIN PLANNING. This instruction caused the computer to take information located at some place in its memory, sometimes called a *directory* or an *account*, and made it available to Junior Planner. In this case, the directory is called PLANNING, and within this location files are stored containing information the planner may want to access, such as GARBPICKUP. There is also enough storage space in the directory for the user to add new files or programs to be used at some other time. By typing the internal command SLIST GARBPICKUP, Junior Planner instructed the computer to display the file, line by line, on the terminal. All computers have these types of internal commands, which can be used for file manipulation, program running, and operating peripheral equipment.

For the user whose main interest in computers is data storage and retrieval, interactive internal commands will be his or her primary concern. Each computer manufacturer has its own internal command structure, and all do about the same kind of operations. Some computers are initiated with the command LOGIN, others by SIGNON, and still others by just turning on the switch. No matter what gets the computer going, good documentation of these commands is usually supplied by the manufacturer.

The computer is a simple machine that must be instructed in exact terms and will respond only to those commands that it has been programmed to accept. Any other message, especially some random collection of characters and numbers not in the computer's internal command language, will not be processed. Generally it will display some statement to that effect, such as NOT A LEGAL COMMAND.

Programming

A program is a collection of logical commands instructing the computer to do some operations. Various computer languages can be used to write the commands, but only a few have become widespread and available on almost all computers currently being marketed. Fortran, BASIC, and Pascal are three of the more common high-level languages available. These are capable of doing the kinds of operations planners may be interested in. Fortran, developed by IBM, was one of the first scientific programming languages and consequently is well known. BASIC, which is similar to Fortran, was developed at Dartmouth College primarily as an instructional programming language designed to teach people how to write computer programs. BASIC has become the language of the personal computers and is one of the most versatile languages available. Most programmers can program in BASIC, as well as some other languages. Pascal, developed by Kathleen Jensen and Niklaus Wirth, is another versatile language currently being distributed by the University of California, San Diego. Pascal is taught extensively on the West Coast and is rapidly becoming available throughout the country.

Software

It is a waste of one's time to write new programs to solve problems that have already been programmed. A planner doing some statistical analyses could, if given enough time, write a statistical programming package that will solve a variety of problems, but it would be far wiser, cheaper, and less time-consuming to buy a program already written by a professional programmer that probably can do even a larger variety of statistical analyses appropriate for the planner's use. These programs, sometimes called *canned programs* but more generally referred to as *software*, are widely available to users of small computers. Most large computers have an extensive library of canned programs. Some programs, like statistical packages, linear programming packages, and financial packages, are used primarily for numerical analyses and are quite reasonably available for most computing systems. Canned programs that can be used to generate graphical displays on terminals, printers, plotters, and the like are also available. In some computers these types of programs can be used ("called up") through statements that the user writes into a program. Finally, canned programs that allow one to do text editing or word processing are available as software packages on many computers.

Although planners should be able to write their own programs, good software should also be part of their computing system. The variety and sophistication of software available to computer users, regardless of the computer's size, increases daily, and planners should take advantage of it. Furthermore, as planners develop programs of specific interest to other planners, publication of those programs in the form of software is advantageous to the profession.

USES OF COMPUTERS IN PLANNING

Most planners use computers to perform three major operations, or some combination of these: arithmetic calculations (technical and economic analyses), data storage and retrieval, and text and graphical presentations.

Arithmetic Calculations: Technical and Economic Analyses

Hand-held calculators have taken over the role of the slide rule. Most of the common arithmetic and trigonometric operations can be performed on a small hand calculator as a result of advanced electronic, or computer, technology. Addition, subtraction, multiplication, and division, accurate to eight places on many hand calculators, are the common arithmetic operations. Sine, cosine, tangent, and their inverse functions are also available on many models. Some hand-held calculators can do considerably more; they can be programmed just like any large computer either to solve some complex arithmetic operation or to perform the same operation for different values of the variables many times over.

The primary use of the early computers was to perform long, tedious calculations for many different values in as short a time as possible. Before computers were widely available, mathematicians used cumbersome, slow, mechanical calculating machines. With the arrival of high-speed computers, complex numerical operations became more common.

Of the many technical areas that are of direct concern to the planner, resource planning and transportation seem to be the most obvious. Because of the technical aspects of these areas, they are most amenable to mathematically quantifiable analyses, and the computer is ideally suited to these areas.

Resource planning

The primary concern of some planners is resource allocation and distribution, and the resources that these planners most frequently deal with are land and water. Recently energy resources have become a major concern to a number of planners, and in the near future, the availability of food, mineral, and metallic resources will affect planning as well.

For the most part resource planners are concerned with the allocation of scarce resources either for an appropriate use, such as land, or to the appropriate user, as with water. A planner needing to know how much water will be available to a particular area for future growth considerations could use the available mathematical models (equations) in a computer program that predicts water availability as a function of land use. The planner then can consider a variety of future growth patterns, which will in turn determine how runoff, stream flow, underground pressure, and similiar parameters will affect future water availability. Either simple or extremely complex models can be used, depending on the accuracy needed and the level of decision making that is ultimately affected.

Similarly the availability of energy resources can be predicted based on estimates of future resource availability under a variety of conditions. For example, a regional energy problem is addressed in the Hiatonka and East Victoria case study. Deregulation, it was projected, would greatly increase the cost of oil and natural gas, and so a study was requested of the use of alternative sources of energy. It was hoped that taking advantage of renewable energy sources in some way would offset the high cost of oil and gas. Technical data dealing with the region's geographical and climatological characteristics were obtained to predict the amount of solar, wind, geothermal, biomass, hydro, and other alternative energy that is available to the area. The computer can be used for this analysis. A plan would be developed from this information to expedite the use of the more abundant and cost-effective alternative energy sources for those uses normally served by oil and natural gas. Either some simple mathematical models could be used, or complex systems analysis could be performed by the computer to analyze the various energy allocation alternatives.

Transportation

Optimizing traffic flow through cities or regions was one of the earlier traffic engineers' computer problems, and now it is the transportation planner's problem. A number of mathematical models evolving out of these transportation studies are now used in many other fields, based upon the premise that the optimal transportation pattern minimizes either the travel distance or travel time or both. Although minimum time and distance traveled are not always valid optimizing criteria when planning traffic flow, they are still a major concern to transportation planners and are ideally suited to computer analysis. The second case study in this book is about a conflict surrounding a transportation study for the medium-size city, Opportune. Besides using the computer for optimally locating transportation routes, the computer could be used to analyze transportation needs. In the Opportune case, it may be more important to show that a particular highway route being mentioned would reduce neither travel time nor distance traveled for the people in the service area. In this way a conflict could be reduced by eliminating those routes that may have been inaccurately mentioned as candidates for construction.

Technical analyses in the area of resource planning and transportation usually fall into two categories: linear programming and statistics. Although a broad collection of sophisticated mathematical techniques is available to computer users, planners will find that most of their complex technical analyses can be accomplished with linear programming and statistical methods.

Economic analyses

Planners at one time or another will perform some kind of economic analysis. It may be a simple analysis, such as picking out the cheapest garbage can from a catalog, or it may be a complex study of the impact a city's budget would have on various segments of the population over a variety of demographic characteristics. However, most planners will probably do only some life-cycle costing, or cost-benefit analysis, or budget preparation and projections. Each of the three case studies described in this book requires some kind of economic analysis.

Computer programs capable of doing these analyses are common. Complex and simple econometric models have been on computers since quantitative economics began to flourish in the early 1960s. Today's planner with a reasonable understanding of econometric modeling is well aware that a large variety of economic policy analyses and economic forecasting programs is available. Forecasts of industrial growth, employment projections, and associated revenues can be made for a city, state, or region based upon well-known input/output models. Data appropriate to the geographical area being analyzed are fed to the computer, analyses are performed, and information is produced for planning purposes in the same manner that other forecasting and impact analyses are used. One must be careful that the model selected is appropriate to the case being analyzed. If the assumptions are not valid or the data are incomplete, then the output and subsequent conclusions are useless and potentially dangerous.

In general, economic analyses done by computers are at the same mathematical level as technical analyses. Linear programming and statistics are the primary analytical tools, but users can create their own programs using other techniques appropriate to economic problems.

Most people are aware that the computer can do arithmetic with great speed and accuracy. What some may not realize is that one does not need an extensive mathematics background to get the computer to do those calculations.

Data Storage and Retrieval

Probably the most common use of computers in large organizations is to store and retrieve vast amounts of information. It is possible to locate one name and address out of hundreds of thousands or one social security number out of millions and display them in a fraction of a second. More than likely, one would be interested in getting a list of all the names and addresses of people in a particular precinct, or all the names of people over sixty-five in a ward, and so on. As the computer's ability to analyze information progressed technically, so did the materials that store the information, in the form of electronic signals. Magnetic tape became widely available for computer usage in the 1950s, and by the early 1960s, magnetic disks were widely available. The electronic chip is also capable of storing large amounts of information in a very small place, and it is this ability to locate information rapidly and quickly from a spinning disk, a rapidly moving tape, or a computer chip that allows one to take advantage of the computer's manipulating capabilities for information storage and retrieval.

As computer technology became more sophisticated, so did the resulting information and the devices on which the information is stored. More and more organizations began to use computers primarily for their data-accessing capabilities. Most organizations have learned that sometimes it is unwise to adopt a complex data-accessing system for a relatively simple data analysis. But this profusion of data-accessing availability to almost anybody who wanted it, rapidly and cheaply, caused the communication age to boom.

Frequently planners need access to large amounts of data, but without the need for sophisticated mathematical analyses. They may simply want a listing of a city's garbage pickup schedule, or a listing of all the commercial buildings over 100,000 square feet of floor area, or any type of detailed information pertinent to a particular problem they might be working on. Since a computer can store, retrieve, and display large amounts of information in a relatively short period of time, it is well suited for data management. Data are not always of a numerical nature. The computer is as adept at sorting or locating letters and words as it is with numbers. For most computers, alphabetizing a list is a straightforward procedure.

In the more complex situations there are large data banks containing many volumes of information in the form of census data, transportation information, or detailed collections of numbers, words and statistics. These data banks may belong to a government agency, a utility, business and industry groups, or any other large organization that either uses the data directly or compiles them for their clients. In many cases a planner can get access to these data banks to use locally. The most general type of access is a listing of either the entire file or, more often, some parts of it. The planner may also request a tape (or disk) containing the data bank that can be used by his or her computer for future data reduction and analysis.

Fortunately for planners, software packages are available that allow one to examine rapidly large amounts of data and present only the information that is of immediate use. These data-base management systems (DBMS) can also format the information in tabular form for cross reference to other key data.

Junior Planner could have used such a system if the garbage pickup data were more detailed and more extensive as they would be for a much larger city. A DBMS would be used to query the data file for all the downtown Main Street pickup dates without having to wait for this information to appear (an inefficient and confusing process for very large data sets). Some of the more sophisticated systems let the user hypothesize specific changes for some of the data and then calculate the effect of these changes on the entire data set. Budgets and schedules, with defined limits, are ideally suited for these types of systems.

Text and Graphical Presentations

The newest surge in computer usage is for text writing and graphical presentations. The traditional office typewriter is rapidly being replaced by a single-purpose computer or one that is part of a larger computing system within an organization. A letter, report, or other small document can be typed, corrected, and even have the spelling checked by a word processor, the term used for such a single-purpose computer. Secretarial typing skills, such as spelling, paper centering, setting margins, and the like, have been greatly enhanced by the word processor. In a large number of organizations new computers used only for their word-processing capabilities have been purchased because of the recent high demand for this usage. Modern high-speed printing devices that produce copy similar to professional printing machines or typewriters are now available for use on almost any kind of word-processing system.

In addition to text writing, most computers can now generate a large variety of graphic displays, such as maps, charts, tables, and figures. Having the tables and charts for a report printed along with the text is a tremendous advantage. The majority of the computer-generated graphics are accomplished with differently sized horizontal, vertical, and 45-degree lines. Occasionally some shading can be simulated by using varying amounts of dots, letters, or numbers on some regions of the figure. This type of shading can be used for certain types of pictorial displays, to represent physical characteristics of a mapped region, or for some shading purpose. More recently high-resolution color graphics have been made available to computer users. Now one can generate multicolor figures, adding a new dimension to computer-generated report preparation and presentation.

The letter, report, or presentation is the planner's final product, and the elaboration of this product can mean the difference between adoption or rejection of his or her recommendations. Computers can print letters with no spelling errors and can also present graphic information in a simple and understandable format.

Not all computer-generated graphics come only from single-purpose word-processing devices. Small, personal computers, or microcomputers, are now frequently used for complex, multicolor computer-generated graphic presentations. As the electronic chips used in these computers become more complex and have greater storage capabilities, the graphic capabilities and display procedures are becoming more elaborate. Planners can use this capability in a variety of ways. For instance, maps can be displayed that show differences in population density, land use, land values, or any other parameter of interest. The use of color to display these changes can have a dramatic effect on the audience when one is trying to present a broad overview of a particular problem. Color radar displays shown on television weather summaries are accomplished by a computer that enhances the radar signal to indicate storms, rain, snow, and other meteorological phenomena in different colors. This type of display is useful in many presentations because of the ease in which dynamic changes can be entered, which in turn leads to a new display. A time-series projection demonstrating changes in land use can be quite dramatic when properly done; projecting future revenues through an analysis of land-use changes could be another dramatic presentation. This method of graphical presentation can easily be displayed on a number of color television monitors, allowing large audiences to view the presentation.

As the use of computers increases, various applications will arise that depend upon the availability of the three capabilities (arithmetic analysis, data management, and text and graphic) to planners. Computer-assisted public participation currently is an area of planner exploration and experimentation that fits that application need. Computers are used for the following kinds of public participation activities: conferencing, polling, interactive computer graphics, and simulation/gaming. These activities use one or more of the general computer capabilities.

Computer conferencing uses a system of remote terminals connected to a central computer. It enables participants in dispersed locations to access the same information simultaneously and to communicate directly with each other by typing messages into their terminals, thus overcoming the difficulties of distance and time lag. A sizable number of people can participate simultaneously as long as the appropriate computer equipment is available. As an example, citizens in scattered small towns might review and react to the planning proposals of a regional staff. The participant's messages appear on the display mechanism of the terminals (either a printer or a cathode ray screen), and data to support or counter discussion points can be called up from the data-base component of the system.

Polling or voting may be done from dispersed locations. In the version used at The University of Michigan, developed by Robert Parnes, the participants view a presentation or discussion of issues and then register their reactions (a one- or two-sentence statement) on the computer, which tallies and stores the responses for later display. Anonymity is preserved (if desired), feedback is instantaneous, and participants gain the benefit of knowing others' reactions.

Computer-assisted interactive graphics have the potential for translating highly complex problems, such as the design of a highway route, into more easily understandable and imaginable terms. The key ingredients for such systems to work are a graphics console for the input and display of data, human judgment, and the complex computer programs (software system) that structure the communication. Finally, simulation/gaming, the fourth major area of computer-assisted public involvement, is amenable to computerization.

The key advantages of computer-assisted public involvement systems are their ability to overcome distance and time lag, provide anonymity, and enable participants to draw upon and actively use the computer's data storage, graphic, and arithmetic capabilities. The drawbacks are the complex computer programs and equipment overhead required to establish such systems, the resistance aroused by such high-technology devices, and the generally small and self-selected number of people who choose to get involved. However, the current computer revolution, which is increasingly popularizing computers and making them more accessible, should increase their use in public involvement in planning.

GENERAL TYPES OF COMPUTERS

Three general types of computers are capable of doing the kinds of analyses planners would be interested in: large, high-speed, time-sharing computers; smaller and somewhat slower time-sharing computers, called minicomputers; and personal or microcomputers. The differences among them, besides cost, are the speed with which the analysis is done, the extensiveness of the data that can be analyzed, and the aesthetic quality of the final output. A planner's different needs and different computers available ultimately will determine the use he or she makes of a computer.

Large Time-Sharing Computers

Large computers have the advantages of size and speed if not too many people are trying to use them at the same time, which is what time-sharing is all about. Time-sharing is the ability of a computer to allow access to more than one user at the same time.

Each operation must be done in some logical sequence at a fixed rate. A single operation (adding, subtracting, or storing or retrieving data, for example) takes an extremely small amount of time, but as millions and millions of operations are performed, the time begins to add up. And not all the computer's functions are used at the same time by any one user. As a result, more than one user can share the computer at the same time, so while one part of the computer is doing something for one user, another part is doing something else for another. If many people are using the computer at the same time, the computer may appear to be operating very slowly, but what is happening is that each user is waiting for the use of some part of the computer.

Another not too obvious fact about large computers is that the word *large* really refers to the computer's ability to control and manipulate large amounts of information. This is the computer's internal memory structure and is usually expressed in thousands of bytes (a byte is loosely defined as the amount of storage space needed for one character, number, or symbol) called *kilobytes* or just *K*. A 64K computer is one with 64,000 bytes of internal memory. Two types of memory are commonly referred to: random access memory (RAM), that part of the computer's memory system in which the user can put input and output information; and read only memory (ROM), where the computer stores its operating system's information and which can be used to control the user's information stored in RAM. The internal memory of a computer is not necessarily the same as the total storage, since large amounts of information can be stored on magnetic disks, tapes, cards, or other storage media peripheral to the computer. But the amount that can be taken at any one time from other storage devices and placed in the computer's memory, along with information stored in ROM, is called the main storage. For most large computers, the main storage capability is measured in millions of bytes (megabytes) of storage. Even very small microcomputers can have large peripheral storage capability, but they are greatly limited by the amount of RAM and ROM they can use at any one time.

Mini Time-Sharing Computers

The major differences between a minicomputer and the large computers are main storage size and speed. In fact, a minicomputer used by a small group of individuals would not be all that distinguishable, to the users, from a large computer. The programming languages, system command language for editing, and other file operations appear essentially the same. One does not begin to notice the differences until programs become very complex or large numbers of data files must be used. The computer soon lets the user know that it has run out of space. In addition, the time for the output to be generated can become intolerably long as the number of users increases. Therefore one should be careful before acquiring a minicomputer to be certain that both the number of users and the complexity of the programs and data files will not reach a point where either the waiting time or lack of sufficient output affects usage. If only a small number of users is anticipated and each uses only a reasonable number of data files with their programs, a minicomputer can be as useful to planners as any large, time-sharing computer.

Microcomputers

The newest computer market to emerge, and estimated to be potentially the largest, is for the personal or microcomputer. These computers currently range in size from around 4K to 64K of main storage, and most of them are also capable of supporting a disk or tape operating system. Initially the micro-computer market was for the computer hobbiest—someone who usually had access to a larger computer and wanted one at home. But the general public soon became aware of them and, because of their relatively low cost, the market exploded. From around the mid-1970s, the personal computer market has become a multi-billion dollar industry. All this came about because of the development of the microprocessor, which allowed the placement of all the electronics needed to run the little computers on one printed circuit board about the size of an 8½ X 11 sheet of paper. This advance greatly reduced the cost of the computer and put it in the reach of many homes, small businesses, and classrooms.

The first generation of these small microcomputers (micros) was relatively limited in its ability to do the variety of operations that larger computers do. Today they can do just about anything either a planner or other computer users might be interested in: arithmetic analyses, text editing, and graphics, as well as data storage and retrieval. The amount of data that can be stored or analyzed at one time is considerably less than that which the larger computers can handle, and the various operations are performed much more slowly. Also, they are called personal computers because, under most configurations, only one person can use them at a time.

Nevertheless the advantages are quite substantial. Their relatively low purchasing and operation costs are their most obvious attraction. The vast amounts of software currently available for micros make them substantially more powerful today than when they first appeared. Their most significant advantage to a large organization is that they can be used in consort with the organization's larger time-sharing computers. Most of today's micros can be hooked up directly or by telephone to operate with larger computers as remote terminals, letting the operator perform some preliminary analyses on his or her personal computer and then using it to transfer information to the larger computer for more sophisticated or complex analyses. After the larger computer does its analyses, the microcomputer can then get the information back in a form amenable to its analytical capabilities. This type of information exchange is the main reason that micros have been added to the organization's system. One simple example would be the use of a microcomputer to edit the text of a report. The report is then transferred to a larger computer that controls a multiprintset, high-speed printer. The microcomputer can command the larger computer to have the report printed out by its printer. The report can then be reformatted, some additional data can be inserted, and the changed report can be sent back to the smaller computer, where some new analysis is performed. In this way the normally expensive, time-consuming text-editing and data-analysis procedures are performed on the cheaper microcomputer, whereas the more complex but speedier data handling and peripheral equipment management functions are done by the larger computer.

Nongeneral Computers

Programmable calculators have some computer capabilities. Most people can quickly learn to program these to do some elementary statistics, financing and interest calculations, and a large variety of other functions. Many programmable calculators have statistical and finance operations already built in, and some of the more advanced ones are almost as sophisticated as microcomputers. They can operate printers that give a hard copy (a permanent record) of the output, and a number of these advanced calculators can use words (letters) as well as numbers in their programs. Although programmable calculators are portable and easy to operate and some even have plug-in modules for additional programming or storage purposes, they are not substantially less expensive than microcomputers, which have greater computing capability. However, if one needs a truly portable computer for work at a remote site, then programmable calculators are the answer. As yet programmable calculators cannot store large amounts of data nor can they produce high-quality text or graphics, but they do have their place for some planning applications. Some initial analyses of data obtained in the field could be an appropriate use of hand-held calculators.

COMPUTER PROGRAMMING AND OPERATION CHARACTERISTICS

It is not necessary to know how to program a computer to use one, but some knowledge of the reasons why it performs in certain ways is helpful in making use of the machine and is useful in communicating with a programmer.

Writing a program

Regardless of which language, or languages, one decides to learn, certain common characteristics become obvious to the user because all computers use essentially the same logical processes. For example, suppose one wanted a program that could take some numerical data, substitute them into an equation, and print out the answer. Specifically, let us suppose one wanted the straight-line distance between any two points on a grid map for some city (see Figure 7.1). The equation for the straight-line distance between two points, *A* and *B*, is

$$D_{AB} = \sqrt{(X_A - X_B)^2 + (Y_A - Y_B)^2}$$

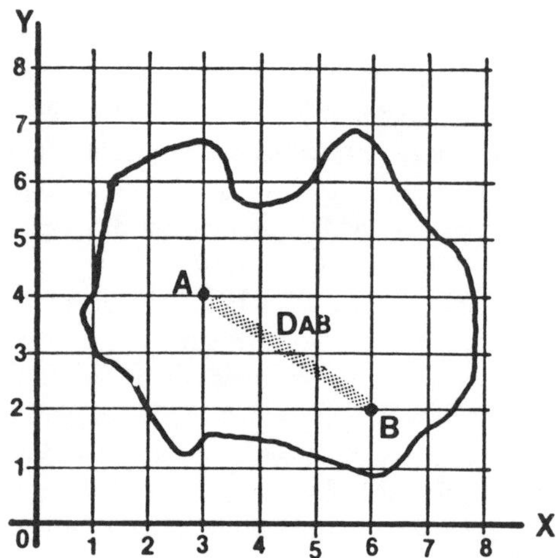

Figure 7.1 Computer map.

where the *X* and *Y* values for points *A* and *B* are taken directly from the map. A simple computer program used to compute D_{AB} for any two points could be the following:

```
1 INPUT XA, XB, YA, YB
2 X1 = (XA – XB)^2
3 Y1 = (YA – YB)^2
4 DAB = SQRT (X1 + Y1)
5 PRINT DAB
6 GO TO 1
7 END
```

The program instructs the user first to enter the numerical values for *XA, XB, YA,* and *YB*. This would mean typing on the terminal the following numbers: 3, 6, 4, 2. The computer calculates from these numbers:

$$X1 = (3 - 6)^2 = (-3)^2 = 9$$
$$Y1 = (4 - 2)^2 = (2)^2 = 4$$
$$DAB = \sqrt{9 + 4} = \sqrt{13} = 3.605513$$

and displays on the terminal the number 3.605513. Then the computer returns to statement one (GO TO 1) and waits for the user to enter four more numbers.

This simple example demonstrates some of the basic processes used in programming. Each statement is unambiguous and must logically follow the previous statement. For instance, if statement 5 (Print Dab) were put between statements 1 and 2, the computer would not be able to follow the logic of that command since the value of *DAB* has not been computed yet.

Another characteristic is that the statements are similar to standard arithmetic, or English abbreviations. For instance,

$$X1 = (XA - XB)\hat{}2$$

is the same as

$$X1 = (X_A - X_B)^2$$

because the computer does not recognize subscripts or superscripts in a program. Similarly, for

$$DAB = \text{SQRT}\,(X1 + Y1)$$

the letters SQRT stand for the square root symbol.

Programming can be relatively straightforward. Even though the example presented here is greatly simplified and some differences between programming arithmetic and standard arithmetic are not discussed, it provides a general idea of what programming is about. With a little practice, one can learn a programming language such as BASIC in a relatively short time. A reasonable amount of practice enables one to develop some sophisticated programs that can calculate numbers from complex equations and produce accompanying graphics. For instance, in the preceding example, picture the map of the city, shown in Figure 7.1, first appearing on the screen (without the points A and B on it). Then picture the user taking a light pen (something that looks like a fat ballpoint pen with a wire attached), touching it to the screen first at point *A* and then at point *B*, and then seeing the value 3.605513 appear on the screen next to a line that also appears on the screen connecting points *A* and *B*. That type of program is well within the capability of most computers and can be written in BASIC language by a moderately experienced programmer.

How They Work

Figure 7.2 is a generalized diagram of a computer operating system. At the left of the diagram, some of the more common input devices, such as a terminal, typewriter, or punched cards, take information and put it into the computer at the input buffer. When enough information is in the buffer for processing, a control system, which is part of the central processing unit (CPU), transfers it to the computer's memory. The information is processed by the CPU using other information the computer has stored in its memory, and then the processed information is transferred from the memory by the control system to the output buffer. Once again it is stored in the buffer until it is ready to be displayed on some output device, such as a terminal, paper printout, or punched cards.

One simple way of understanding how the processor operates on information is to think of it in terms of a group of switches arranged in an orderly fashion. If a switch is in the "on" position, an electric current can pass through and some electronic device can detect that current. If the switch is off, no current flows, and some device will detect that as well. By assigning a numerical value of one to a switch that is on and zero to a switch that is off, the computer can generate any number by keeping track of which switches are on and which are off. Readers familiar with binary arithmetic know what number is represented by the following sequence of six on and off switches: 100111. That sequence corresponds to the number 39 (32+0+0+4+2+1). Any number can be represented by some unique sequence of zeros and ones, just as the number 39 in binary can be represented only by the sequence 100111. In computer language, the zeros and ones are referred to as *bits*. In the example, it took six bits to represent the number 39, and the largest number one could represent with six bits is 63 (32+16+8+4+2+1). The computer has logic circuits that sense some fixed number of bits at any one time, and the smallest sequence that can be sensed, or addressed, is called a *byte*. For most computers a byte is eight bits, and any number, letter, or symbol can be represented by a unique sequence of bytes.

In addition to the thousands of bytes that go into a computer, a computer needs some type of logic circuitry to distinguish among numbers, letters, and symbols. The circuits must also determine what the computer should do with the information once it is processed. For the most part the computer simply adds or subtracts in binary arithmetic. The computer's logic circuits sense which switches are on and which are off, and these circuits turn some other switches on or off and store the results in some appropriate place. Multiplication and division can be accomplished by a series of additions or subtractions. Letters and symbols are numerically coded and flagged as

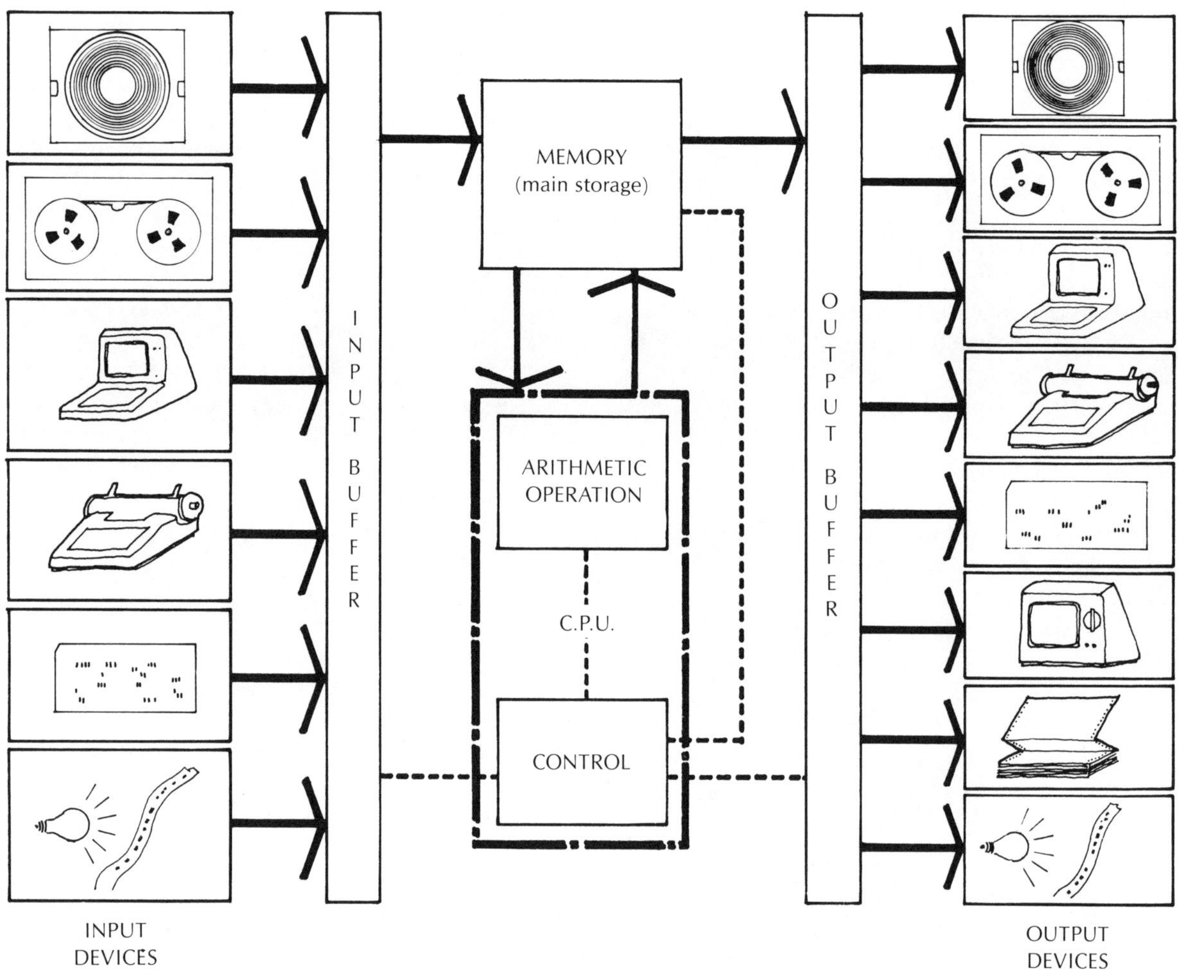

Figure 7.2 Generalized computer operation diagram.

nonnumbers by the computer, using some special bits for just that purpose.

It is important to note some cautions. First of all, computers do "go down"; in other words, they stop working. If one is planning on using a computer system for a demonstration or presentation at some specific time, one should plan on the contingency that the computer will be down at the exact time it is needed. Computers also make mistakes. Granted, most of the time it is the program that is in error, but on some rare occasions, wrong numbers, wrong words or letters, or both, are displayed through no fault of the user or programmer. Computers can lose information stored internally, so one should have a duplicate of any file going into the computer stored elsewhere.

SUMMARY

The use of computers in planning is at a threshold primarily because of the recent availability of reasonably priced computers and the extensive amount of software development. There is little doubt that planners will be using some form of computers for processing the types of information discussed in this book: analyzing field and survey data, accessing large data banks and information sources, performing technical analysis, assisting public involvement by easier participation, generating written reports and enhancing graphic capabilities, and augmenting oral presentations.

This threshold is an important one because there is always the danger of becoming overcomputerized. Generating more data than necessary, performing more detailed analysis than the data are suited for, or simply using the computer because it is available are all bound to happen as the computer is absorbed into the planning process. Each planner will reach his or her own level of computer usage as it fits each particular need.

APPLICATIONS

1. As a planner for the city of Middlesville, you are asked to present various scenarios of how the city might look ten years after the approval of the zoning change. List some of the ways a computer can assist you in developing and presenting your scenarios.
2. In Chapter 2 survey methods are described and questionnaire design is discussed in detail. Describe how a computer can be used to assist in analyzing the results of a survey and what advantages this may have in the selection of your survey questions.
3. Contact a private and a public planning agency and inquire about their current computer usage. Determine their projected future computer usage and how they expect to meet their needs. What differences, if any, are there between the two types of agencies?
4. You have just been hired as a planner for a small planning firm specializing in environmental, transportation, and energy planning. Its major clients are small businesses and cities of under 100,000 population. You are asked to evaluate computer systems for the firm. What capabilities should you look for in a system?

BIBLIOGRAPHY

Most of the many books on computers are addressed to the computer designer or programmer. Since planning does not have an extensive history of computer applications, trying to single out any one or two items would not be of value to the reader. The reader should investigate his or her own computer needs based on the expertise of an adviser capable of evaluating the current state of computer technology.

For the interested novice, a number of popular microcomputer magazines are available that are written for the computer user. Three of the more popular ones are:

BYTE, BYTE Publications Inc., 70 Main Street, Peterborough, New Hampshire 03458.

Creative Computing, Creative Computing, P.O. Box 789-M, Morristown, New Jersey 07960.

Microcomputer, Wayne Green, Inc., 80 Pine St., Peterborough, New Hampshire 03458.

In addition, see *Planning* **47**(10) (October 1981). This issue is devoted exclusively to the use of computers in planning and contains some valuable information on the acquisition and use of computers. An entire issue of *Science*, **251**(4534) (Feb. 12, 1982), is also devoted to the computer and its impact on science and society. (*Science* is published by the American Association for the Advancement of Science, 1515 Massachusetts Ave., N.W., Washington, D.C. 20005.)

PART III

INFORMATION COMMUNICATION

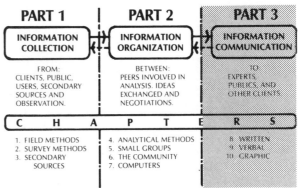

PART 1	PART 2	PART 3
INFORMATION COLLECTION	INFORMATION ORGANIZATION	INFORMATION COMMUNICATION
FROM: CLIENTS, PUBLIC, USERS, SECONDARY SOURCES AND OBSERVATION.	BETWEEN: PEERS INVOLVED IN ANALYSIS. IDEAS EXCHANGED AND NEGOTIATIONS.	TO: EXPERTS, PUBLICS, AND OTHER CLIENTS.
C H A	P T E	R S
1. FIELD METHODS 2. SURVEY METHODS 3. SECONDARY SOURCES	4. ANALYTICAL METHODS 5. SMALL GROUPS 6. THE COMMUNITY 7. COMPUTERS	8. WRITTEN 9. VERBAL 10. GRAPHIC

Effective communication is central to successful planning practice. Interpersonal, verbal, written, and graphic skills are all required. The first of these was already discussed in Chapter 5. The chapters in this part discuss the others.

8

Considerations for Verbal Presentations

Alfred W. Storey

It is important for planners to have effective verbal communication skills. The settings vary in which they must use their verbal skills. They may conduct field interviews, participate as consultants to subcommittees, make verbal presentations to planning commissions and city councils, and speak to large public gatherings.

Since planning is an interdisciplinary field, planners often work in large groups or team settings. It is essential for planners to be able to articulate their ideas clearly to their colleagues in group and team settings. They need to be able to participate effectively as their groups or teams make verbal presentations. If planners want to be effective in these settings and get others to accept their ideas and proposals, they must develop verbal skills which will enable them to communicate effectively.

PREPARATION FOR VERBAL PRESENTATION

A number of factors govern the preparation of a good verbal presentation: the speaker, the purpose, the time allotted for speaking, the audience, and the occasion. Junior Planner took some of these factors into account when he prepared for his first meeting with members of the garden club. When he spoke to them, he told them how impressed the mayor was with their visit, how interested the mayor and city planner were about their concerns regarding trash pickup, and how his task was to find a satisfactory solution to the problem. He also gave attention to audience and purpose factors by studying whom he would be meeting with, by knowing the names of their delegation to the mayor, and by gathering data through on-site visits, interviews, observations, pictures, and cost estimates, so that an equitable solution could be found for the problem of trash buildup on Main Street.

The Speaker

It is assumed that the planner who is called upon to make a verbal presentation knows what issues need to be addressed and is knowledgeable about the material. The planner should be both authoritative and enthusiastic in the presentation, demonstrating expertise and projecting enthusiasm without becoming overly technical. It is important to find the right balance.

Some tension, anxiety, or nervousness before speaking to an audience is natural, but there are ways to control it.

1. Recognize that some disturbance of our mental, physical, and emotional balance is a normal accompaniment of the speaking situation. The pounding of the heart, the feeling of dryness around the lips, the sweaty palms, the sinking feeling in the stomach, and the quivering knees are signs that we are getting ready to function at a higher level of efficiency. Although the speaker is aware of these signs of stage fright, audience members seldom notice them. Knowing this, we will gain more confidence in ourselves and will learn how to control these feelings.

2. Be well prepared. This is one of the most important deterrents to stage fright. Knowing that the speech to be presented is good and that the audience will gain something from listening helps one's confidence. Having a carefully prepared outline and writing out the first sentence of the speech will minimize feelings of uncertainty.

3. The speaker should keep his mind on the audience, not on himself. He should think of how the speech will help the audience. The speaker should approach the speech with the idea that he is going to converse with members of the audience to influence them to take an action that will be for their own good. There should be no fear if the speaker knows that he is trying to meet a need of the audience and that what he has to say is of value.

Purpose

Every speech must have a purpose, and all the material included in the speech must be judged on the basis of whether it contributes to achieving the purpose. The purpose must be attainable within the constraints of the time, the audience, and the occasion.

Usually the planner will have one or more of the following purposes in mind: to actuate (to get the audience to take a particular action), to convince (to bring the audience to a particular point of view), or to instruct (to give the audience particular information about a specified item). The planner should develop material for the presentation to support the appropriate purpose.

Time

The planner needs to know how much time is allotted for the presentation in order to select material that can be treated adequately within the time limit. A great deal of effectiveness may be lost by a speaker if the allotted time is overrun or ignored. Audience members tend to become distracted, tired, fidgety, or bored, especially if several speakers have talked on the same subject.

Audience

The speaker should keep in mind a number of factors about the audience for any given speech. For example, it makes a difference whether a speaker is to address members of a city council or members of a city planning commission. If the topic is rehabilitating or restoring a particular group of buildings in a section of the city, the planning commission members may well be more attuned to the speaker's content, language, and objectives. Council members, on the other hand, would have a broader responsibility for city affairs and would be less able to understand technical jargon and specific points related to planning.

Some of the factors that must be included in the speaker's consideration of the audience are size, age, knowledge of the subject, education, attitude toward speaker and purpose, position on the issue being discussed, economic status, and political views. Many of these factors are present in the case study relating to the application for a required zoning change. A planner making a presentation in that situation might have a wide age differential in the audience because it would include both senior citizens and members of the chamber of commerce. Different vested interests—members of a newly formed neighborhood coalition, of two minority groups, and of the downtown business association—would be in the audience.

Occasion

A planner must know the nature of the occasion for the oral presentation. This information helps the planner to know whether to prepare material to actuate, to convince, or to instruct the audience. It is also necessary to know the makeup of the audience, why the audience is there, who the other speakers are on the program, and the physical conditions surrounding the occasion—that is, whether the speech is to be delivered indoors or out of doors, the nature of the acoustics, and whether a public address system is necessary.

Knowing the nature of the occasion also helps the planner formulate a strategy for identifying and dealing with the important issues. One strategy is to make certain that available information is made known to the listeners, that the pros and cons of issues are set forth, and that various options for solutions are considered. This strategy is based on a well-known procedure for solving problems called the *reflective thought process.* An outline for a speech using this strategy would include four steps, in this order: definition of the problem or issues, analysis of evidence and information, consideration of possible solutions (pros and cons), and selection of final solution.

MODE OF VERBAL DELIVERY

Two common forms for delivering speeches are the manuscript speech and the extemporaneous speech. The manuscript speech is written out in its entirety and read from the printed page. It is perhaps the most difficult type of speech to deliver effectively to a visible audience. It is useful in a situation in which the speaker must have accuracy, perhaps use statistical data, and in which the audience is not visible, such as on radio.

The extemporaneous speech is the form most useful to planners. In the extempore style of speaking, generally a word or sentence outline helps to give careful attention to the organization and amount of material to be presented in a given time. The speech is not written out in paragraphs and sentences; rather, words are chosen at the time of delivery, so careful preparation is essential.

The extemporaneous style of speaking offers several advantages. The speaker can be more direct with the audience in the use of voice, physical appearance, and gestures. Eye contact can be maintained with the audience more easily, and the speaker can determine whether the audience appears to be listening. The speaker can use charts, drawings, maps, and other forms of audiovisual material without diverting his or her attention from the audience to a manuscript. And the speaker will be able to make verbal adjustments as he or she moves along in the speech because he or she is not tied to a manuscript.

MAKING AN OUTLINE

A speech outline can help the speaker in several ways. It can help to organize his or her thoughts and material logically and sequentially, to develop the proper amount of information within the time frame allowed, to present the ideas in an orderly fashion, with reminders for appropriate emphasis and impact, and to prevent the anguish that could come if he or she were to have memory lapses under the pressure of the moment. The outline is a graphic picture of what the speaker plans to say and how he or she plans to say it. The outline provides a thumbnail sketch of the entire speech. It serves as the framework on which the speech is built, forces the speaker to think through the ideas with some care, tends to ensure unity, keeping out irrelevant materials, and serves as a memory aid in the actual speaking situation.

The outline is based on a few simple rules.

1. Each unit in the outline should contain only one item or statement.
2. The items in it should be logically subordinated.
3. The logical relation of the items included should be shown by proper indentation and proper symbols.
4. The organization of the outline should be simple and clear.

The key word outline, which uses only sentence fragments, and the complete sentence outline, which uses only complete sentences, are the two kinds of outlines that are best suited for preparation of an extemporaneous speech. The key word outline is useful for the speaker who needs a guideline or must use notes. The key words or phrases provide a reminder of the order of the material but allow the speaker to choose words during the speech. The key word outline need not appear sensible to anyone other than the speaker. The complete sentence outline forces a clear organization of the materials, the proper integration and subordination of all the ideas to be presented, and a complete preparation of the speech.

Let us assume that you, as a planner, are going to make a presentation to the Middlesville Planning Commission on the matter of a zoning change as suggested in the first case study. You might prepare this key word outline:

I. Urban Blight
  A. Broken roads
  B. Deteriorated apartment dwellings
  C. Boarded-up storefronts
  D. Congested parking and abandoned cars
II. Things to Change
  A. Repair streets
  B. Renovate apartment structures
  C. Repair and reopen storefronts
  D. Paint parking lanes; remove abandoned cars
III. Support and Coordination
  A. Chamber of commerce and city council
  B. Small business loans
  C. Urban Renewal Grant—federal government

A complete sentence outline for the same occasion might look like this:

I. As Middlesville grew older certain sections of the city were neglected. The rundown sections of the city led to a condition called *urban blight*.
  A. Streets are in need of repair, with numerous potholes and broken curbs.
  B. Apartment buildings are deteriorating. The mortar is falling away from brick walls, paint is peeling off wooden structures and fire escapes, and lawns are overrun with weeds.
  C. Some businesses have closed in the area. Some storefronts are boarded up, and others have the word *closed* written on their windows, adding to the atmosphere of decay and deterioration.
  D. Parking lines have faded, and old cars have been abandoned, leaving streets in a congested condition.
II. The following changes need to be made.
  A. The streets should be paved with a new layer of asphalt, and whole sections of curbing should be replaced.
  B. Mortar in old brick apartment buildings should be replaced, and wooden siding, trim, and fire escapes should be painted. Weed and debris should be removed from yards around the buildings, and lawns and shrubbery should be trimmed.
  C. Boards should be removed from storefronts, and owners of vacant stores or buildings should be required to maintain the stores and buildings in good condition.
  D. Parking lanes should be repainted, and abandoned cars should be removed from the streets.

III. Financial and political support and coordination are needed to rid the city of urban blight.
   A. The chamber of commerce should fill a vital role and coordinate the efforts of the business community, local government, and the public in this urban renewal effort. The city council should give leadership by providing for the street improvements.
   B. The city council and owners of small businesses should work together to secure small business loans for refurbishing and reopening some of the stores.
   C. A coordinated effort should be made to secure an Urban Renewal Grant from the federal government.

WAYS TO ORGANIZE THE PRESENTATION

The organization of a speech is an important element in the oral communication process. The speaker must organize the thoughts and ideas to be presented in a meaningful pattern.

Probably there is no single, best way to organize a speech; however, the basic pattern that serves as a model for organizing most types of speeches is the three-part organization: introduction, body or discussion, and conclusion.

Introduction

The introduction should be prepared last. Only after the conclusion and the body or discussion have been worked out will the speaker know what to incorporate in the introduction. Often first impressions are strong impressions so what the speaker does at the beginning of the speech is important to the overall success of the speaker's mission. Four important functions of the introduction are: to gain the attention of the audience, to arouse the interest of the audience, to present the theme of the central idea of the speech, and to suggest (or to state) the purpose of the speech.

The first part of the introduction of any speech has the purpose of gaining the attention and arousing the interest of the audience. There are at least five ways of opening the speech: by referring to circumstances attending the speaker's appearance in front of the audience, by telling a story (serious or humorous), by alluding to a timely and striking statement, by alluding to a timely and important incident, or by asking a question. To illustrate how these five ways of opening a speech might work, let us consider how Junior Planner might have begun his first verbal presentation to the garden club.

Reference to attendant circumstances

"Credit for what I have to say goes to Mayor Lorch and to you who went to see the Mayor—Ms. B., Mr. Y., Mr. K., and Ms. L. You have brought an important matter regarding the health and appearance of our city to the mayor's attention. The mayor is in complete agreement with you and has asked the city planner's office to work with your organization to find the best possible solution to the problem."

Story (narration and description)

"It is a beautiful, spring, Monday morning in Middlesville. On this day our town is to be visited by the state's Beautification Awards Committee. The car bearing the four committee members enters town from the north, crosses over our beautiful, winding river and moves along the tree-lined Elm Street, with its well-cared-for homes and neatly trimmed lawns. As the car turns from Elm Street onto Main Street, the first sight to greet the committee's eyes is . . ."

Allusion to a timely and striking statement

"'If people of Middlesville will channel their requests for city improvement through the system, the system will be responsive.' Have you ever heard this statement? I am meeting with you today because Mayor Lorch and the city planning director believe . . ."

Allusion to a timely and important incident

"On Friday, the day after the city planning director asked me to meet with you about the collection of trash on Main Street, the mayor's office received word that the state's Beautification Awards Committee is scheduled to visit Middlesville some time next month. The timeliness of your visit to the mayor's office . . ."

A question

"Do you believe in the old saying, 'There is nothing that can't be done that can't be done!' I do. That is why the image of Middlesville . . ."

The closing part of the introduction should include the central idea of the speech. This is accomplished by making a direct statement or by asking one or more questions that suggest the theme. In this way, the purpose of the speech is given and the introduction is drawn to a close. By using one of the types of introductions, the speaker should be able to gain the attention and arouse the interest of the audience.

Body or Discussion

The body or discussion is the heart of the speech. This is where the speaker does what the introduction says will be done. In the body the main ideas are developed, the analysis takes place, the evidence is presented, and the major statements and their subordinate points are presented. The length of time for the speech will determine the extent of the treatment of the development of the idea.

Four ways a speaker may develop an idea in the body or discussion of a speech are: exposition, argumentation, description, or narration. Exposition is explanation. Argumentation defends one side of a proposition. Description tells how a particular point of view, a product, an object, a person, or a situation appears to the senses—that is, how it looks, tastes, feels, smells, or sounds. Narration tells a story.

To illustrate, in exposition a speaker explains the rules of the game of basketball or how to work a certain play. In argumentation the speaker attempts to prove that one basketball team is better than another. In description, the speaker gives some of the details surrounding a tournament game that appeal to the senses: the attitude of the crowd, the activities of the cheerleaders, the colorful flags and banners, the pep band, and the whistle of the game officials. In narration, the speaker tells how a particular game was played from the opening center jump to the final whistle.

Probably the most useful ways for the planner are exposition and argumentation. Whatever form is used, the speaker fills in the details. He or she must decide whether to use logical and/or emotional appeals and must decide what form to follow depending on the occasion, the audience, the purpose of the speech, and other parameters.

It is in the body or discussion of the speech that one should be able to determine the speaker's train of thought. Therefore, the speaker must have that train of thought well organized and well supported with illustrations, anecdotes, reasoning, emotionally evocative material, and evidence to guide the audience.

Conclusion

After the body or discussion of the speech has been developed, it should be rounded out, given a note of finality, and brought to a conclusion. The conclusion should not be a perfunctory, tacked-on addition but should serve as the climax of the speech. The conclusion is important because it represents the last opportunity the speaker has to make a point clear to the audience. The conclusion should be developed so that the speech is not ended abruptly, and the conclusion should provide the speaker with an additional opportunity to reach the audience. Special care should be given to the conclusion, for it is the speaker's last chance with the audience. The conclusion should be brief and simple and have both unity and energy.

In general, conclusions are of three types, with respect to purpose: summary, application, and motivation.

Conclusion of summary

Designed to provide a bird's-eye view of the speech, this type of conclusion may consist of a formal summary, a pointed or epigrammatic statement, or an illustration containing the essential ideas of the body or discussion. While a formal summary may be adequate with regard to restatement of material, it may frequently be desirable to restate the substance of the speech in a new way. Hence, not only a brief paraphrase but also a pointed statement or an illustration may be useful.

Conclusion of application

As the speaker develops the central idea of the speech, the listeners may say in effect, "Yes, I see the point, but what should we do about it?" A speaker who expects them to do anything should apply the central idea to the audience by doing one, or both, of two things: apply the theme and its subordinate ideas to the interests of the listeners or suggest procedures available to the listeners. In the first instance, the speaker should relate his or her ideas to the attitudes, ideals, vocations, and avocations of the listeners. In the second instance, the speaker should ask listeners to do such specific things as sign, buy, investigate, vote, donate, write, or whatever is appropriate. The speaker should offer the audience a way to translate belief into action.

Conclusion of motivation

A speaker who wishes to do more than make ideas clear (conclusion of summary) or present means of doing something about it (conclusion of application) should appeal to the basic motives or desires of the listeners. The speaker needs to supplement logical appeal with a psychological appeal to such desires as self-preservation, reputation, affection for others, sentiments, or property. The speaker should relate the central idea to one or more such basic motives, thus tapping the resources inherent in every human being.

SKILLS IN DELIVERY

Three of the most important skills a speaker can develop that will enhance speaking are voice, bodily action, and directness.

Voice

A clear, well-understood voice is a tremendous asset in oral communication. The voice describes the speaker and reflects the speaker's emotions. Often the tone and nature of the voice tells us what an individual is like and whether that person is happy, angry, or bored.

A good voice is one that calls little attention to itself. It promotes the communication of ideas. A disagreeable voice is likely to attract attention to itself and to cause the listener to miss the communication of ideas. A speaker who wishes to have a good voice so that listeners focus on the message rather than on the voice needs to study and perfect at least three essentials of voice control: audibility, quality, and vocal variety.

Audibility

Questions one might ask regarding audibility are: Is there adequate volume? Is the volume adjusted to the room or the situation? It is the speaker's responsibility at the outset to make sure that the volume of his or her voice is adequate to meet the room or situation. There is not much point to presenting a speech if members of the audience cannot hear it. Frequently speakers begin a speech and move well into it before determining whether members of the audience can hear adequately. It is useful for a speaker to select someone in the back row of the audience and ask if that person can hear what is being said. If that person cannot hear, the speaker must make the appropriate adjustment by raising his or her voice or by using a voice amplifier.

Often speakers prefer to speak without the aid of a public address system, but the system is a means of making certain that audience members hear adequately. A speaker should test the distance between his mouth and the microphone to be sure that his voice is heard adequately.

Quality

Each voice has a timbre of its own due to the pitch and resonance used. A voice with a pleasant tone is acceptable to the human ear. A pleasant voice does not call attention to itself, and thus it is easier for listeners to give attention to what is being said rather than how it is being said.

Vocal variety

Of the three essentials of voice control, probably vocal variety is the most important. To produce vocal variety, pitch, rate, and loudness must change according to the mode and meaning of the message. Although a speaker may seem to have little immediate control over the fundamental quality of his or her voice, the control over pitch, rate, and loudness is absolute. In the communication effort, vocal variety will make the voice more interesting to an audience and help hold its attention and give more emphasis and meaning to the speaker's words. Too much variation can take place when pitch, rate, and loudness are changed too rapidly or are not changed for communication purposes.

Pitch in speech may be described as the tone of one's voice. As in music, the pitch would be any given point on the musical scale. In speaking, pitch is the tone, or the point of sound on the speaking scale, in which the voice is heard. The pitch that one has varies with the individual and what nature has provided. It can be controlled with conscious effort. One does not usually enjoy listening to a voice that continues on the same level or same plane. Pitch should be varied with the content of what is being said. A speaker who gives no variety to pitch will not have a natural, pleasing, and impressive voice. If the tone of the voice rises to high, unnatural levels, the pitch becomes unpleasant. A natural pitch for a speaker may be at a high tone level or a low tone level. The best use of pitch demands upward and downward movement, relieving monotony and serving to express and emphasize different thoughts and emotions. The upward and downward movement in pitch is known as *inflection*. Inflection is controllable and, when given conscious effort, helps the intelligibility of thought.

The rate of speech is judged by the number of words uttered in one minute. Researchers have found that the average rate of speaking is 125 to 150 words per minute, although people can understand words at the rate of 250 words per minute. Just as important as whether a person speaks too slowly or too rapidly is the matter of whether a person speaks at the same rate. Most listeners dislike a dragged-out rate of speaking or awkward pausing; they also dislike speech that is so rapid that emphasis is impossible. Therefore it is important to vary one's rate of speech.

The pause is an important element in varying the rate of speech. A variety of emotions will demand a variety in the rate. The length and frequency of pauses are determined by the sentiment expressed. Pauses should be used knowingly to change the rate of speech to suit the speaker's intellectual and emotional modes. If the speaker is concerned with better communication of ideas, then attention should be given to variety in the speaking rate.

A third important component of variety is loudness. Speakers have control over loudness but often do not use that control to advantage. Changes in loudness, or vocal force, may be used to relieve monotony and to secure interest and emphasis in speech. Some voices are too loud, but a far greater number are too weak. Generally if one varies the pitch of the voice appropriately, loudness variations will be adequate. Some people tend to decrease loudness near the end of a phrase or sentence so that a listener fails to hear the last words in the sentence. This is called "dropping" the voice. Sometimes persons associate loudness with emphasis and do not understand that a subdued voice can serve the same purpose.

An increase in loudness does not necessarily mean a higher pitch. Though pitch may be raised with more force, that does not always occur. Loudness may mean that one makes the voice more audible. The voice becomes louder as the need for emphasis or emotion dictates. It must be understood, however, that one may also gain greater emphasis with a decrease in loudness. In both cases, the movement of the voice for greater or less loudness adds variation to speaking. The variation in loudness must be in accordance with the size of the room, the occasion, the size of the audience, and the emotional state of the speaker.

Bodily Action

An important aspect of delivery is bodily action. Does the speaker use his body to aid his speaking? Does the speaker appear to be at ease and physically comfortable? Is the bodily action of the speaker—that is, facial expression, gesture, posture, and general physical movement—helpful to the audience's understanding of the message?

Sometimes speakers inhibit their normal bodily action when speaking to an audience. If one's gestures are not synchronized with what is being said verbally, such behavior will be distracting to an audience and detract from the content of the speech. If a speaker shuffles across the platform and does not appear alert and vital in both posture and bodily movement, audience members will assume that there is lack of conviction, preparation, or commitment to what is being said. Such behavior will detract from the audience's acceptance of the speaker's message.

Effective bodily action promotes the communication process in several ways. It attracts the attention of listeners and helps them maintain an interest in the speaker. The oral performance becomes more attractive and interesting. The meaning of words and ideas is aided by the speaker's visual cues that provide the audience with a means of evaluating the speaker's attitude and intention toward them.

Bodily action serves as a transition aid within the speech and helps to integrate the total speech performance. If bodily action is vigorous and alert, it will help make the voice vigorous and alert.

Most people use facial expression, hand gestures, head movement, and other forms of bodily action when speaking in normal conversation, but they tend to freeze before an audience. Generally experience and coaching help an individual to be as free with body movement when on the platform as when off the platform. The speaker who employs effective bodily action stands a better chance of winning and holding the attention of an audience than the speaker who does not.

Directness

One of the most important aspects of effective speaking is maintaining direct contact with the audience. Probably no single action on the part of a speaker is more important than direct eye contact with the listeners. As a speaker looks directly in the eyes of individuals in an audience, those people in turn feel a greater interest in the speaker and have a desire to continue to look at the speaker. Such attention improves the chances of having the audience listen to and understand more clearly the intent of the speaker's message. Audience members may be distracted by a speaker who stares above their heads or who looks at the floor or ground while speaking. A major reason for speakers to retain a high quality of directness through eye contact with audience members is that it lets the speaker know whether audience members are paying attention, whether what is said appears to be received by the audience, and whether one's remarks are on target.

Direct eye contact greatly aids other forms of directness in a speaker's approach to an audience. For example, it would be virtually impossible for a speaker to look audience members directly in the eye without at the same time using gestures, head movement, and other bodily action that coincide with what is being said vocally and attract and hold their attention. Direct eye contact also helps the speaker know whether his voice is heard by all members of the audience, it fosters direct response from audience members, and it enhances the communication process between the speaker and the audience.

In summary, it is one thing to talk at an audience, another to talk to its individuals. Someone has described the effect that certain public speakers produce as spraying water over a garden with a hose. The words and sentences are sprayed into the air over the audience, falling like drops of water. Public speaking is private speaking magnified and intensified. When two people talk together on a topic of common interest, they do not spray each other with words or talk at each other. They talk genuinely and earnestly to each other. Adopting this attitude is the most basic step in the development of directness.

Some aids for securing directness are:

1. Look directly at the audience and focus on individuals all the time. Actually watch your audience and consciously record its reactions in an attempt to adapt your speech and manner to it. Audiences are as different as individuals and must be observed carefully while you are speaking.
2. Concentrate upon communicating your ideas to the audience. All movement should arise spontaneously from this concentration.
3. Maintain good, alert posture and use gestures. These suggest directness to an audience. Face an audience squarely, walk deliberately, and stand deliberately still when you are not walking. Gestures synchronized with the words give added meaning and emphasis to what you say.
4. Cultivate a pleasant and mobile facial expression. This is an outward expression of enthusiasm and earnest friendliness. Reflect your thinking in the facial expression.

Some hindrances to directness are:

1. Self-absorption. The speaker who worries about himself on the speaking platform cannot communicate directly with the audience.

2. Fear of making mistakes. Theodore Roosevelt said, "Show me the man who does not make mistakes, and I will show you the man who does not do anything." If you make a mistake, continue with the speech.
3. Stage-fright. Some disturbance in the normal mental, physical, and emotional balance is inevitable in the speaking situation. People who have spoken for years still experience stage-fright.

Gestures, bodily action and directness may be improved by following the suggestions in Exercises 1, 4, and 5 at the end of this chapter.

TECHNIQUES OF DEMONSTRATION

Often visual aids help the communication process. They have these important advantages:

1. Clarity. Words may not always have the same meaning to both the speaker and the audience. If listeners can see what the speaker is talking about, as well as hear it, they are likely to understand more fully.
2. Audience attention. Generally an audience will show more interest and attention when visual aids are used. The longer a presentation is, the more important the use of visual aids becomes for the purpose of maintaining the attention of the audience.
3. Memory. An audience is likely to remember more of the material and for a longer period of time, when visual aids are used. The aids also help the speaker remember more clearly what he plans to say.
4. Poise. Handling and pointing to visual aids gives the speaker additional reasons for moving about and for developing poise. Maintaining poise can help ease tension for both the speaker and the audience and can help strengthen the impression the speaker makes upon the audience.

Charts, diagrams, maps, and pictures are important aids in explaining difficult or technical subjects. There are a few simple rules to be followed and some obvious mistakes that can be avoided in the use of visual aids.

1. The charts should be large enough to be seen. If the design on the chart is not large enough to be seen, it will become an annoyance. Do not guess the proper size while making the presentation. Draw an experimental chart or diagram before hand. Test it out in the room in which the presentation is to be made to determine whether the outlines of the chart or diagram can easily be seen in detail. Make appropriate adjustments.

The lines of the chart or diagram should be heavy and broad. Even the writing on a chalkboard should be carefully checked for size and for heaviness of line.

2. Do not crowd too many details into one chart; they can lead to confusion. Avoid needless complexity. If there are several explanations to be made or a series of steps in a process to be explained, do not try to put them all on one diagram. Use a series of diagrams, each as simple as possible.

3. Talk to the audience in strong positive tones, not to the chalkboard. Learn the art of keeping a pointer properly placed on the chalkboard while looking at the audience.

4. Do not stand between the audience and the chart. As a speaker should not ignore an audience and talk to the chalkboard, so the speaker should not ignore the chalkboard while talking to the audience. If the audience is seated very close to the speaker, then the speaker should be certain to stand a bit to one side and, in most cases, use a pointer.

5. Do not let an unused chart distract attention. If possible, charts should be kept out of sight until needed and removed from sight when no longer needed. Audience members will look at a chart or picture as soon as it appears and try to figure out its purpose. If a visual aid is exposed to the audience before it is ready to be used, it may become distracting and prevent the audience from listening as attentively as possible to the speaker.

SUMMARY

Planners need effective verbal communication skills. How they gather and organize material; how they analyze the occasion, the audience, and the purpose of their verbal presentations; and how they develop and use their voices, gestures, bodily action, and eye contact to a large extent can determine the success of their verbal presentations. The goal is to be able to transmit ideas to listeners so that they understand the intent of the message.

We can become aware of our own needs as oral communicators and improve our abilities by applying some of the suggestions offered here. Each of us has developed our own, unique style of speaking, and we can become more effective speakers. We need to understand which parts of our speaking personality need improvement, to know what techniques and skills we can apply to bring about that improvement, and to practice. We can become more effective and successful planners if we improve our verbal communication skills.

FOR FURTHER CONSIDERATION

1. A Checklist to Study
   Rate yourself on the following statements with a 1, 2, 3, 4, or 5:

   1 = Very seldom.
   2 = About 1/4 of the time.
   3 = About 1/2 of the time.
   4 = About 3/4 of the time.
   5 = Almost always.

   My Speaking
   ______ a. When speaking I pronounce words clearly to make it easy for listeners to understand.
   ______ b. When speaking I look directly at members of the audience.
   ______ c. When speaking I do not do other things (arrange papers, play with a pencil or pointer, jingle coins in my pocket) that may distract the attention of my audience.

   My Attitude
   ______ a. I research my topic thoroughly and speak in a confident manner.
   ______ b. I show respect toward the person or persons with whom I am talking.
   ______ c. When I speak to an audience I make certain that I understand them—their frame of reference—their situation and environment.

2. Analyze a popular speaker on television, at a public meeting in your town, or at your church. Ask yourself these questions as you listen to and observe the speaker in order to determine what was the most effective and least effective. Did the speaker:

   a. Use his or her voice to the best advantage?
   b. Vary his or her rate?
   c. Converse with the audience?
   d. Read the speech? Memorize it? Use notes?
   e. Use meaningful gestures?
   f. Use his or her body for communication?
   g. Use effective eye contact?
   h. Include an introduction, body, and conclusion?

3. Prepare a two-minute presentation on a planning project. Imagine an audience of thirty people. Put your speech on a tape recorder and listen to the playback. (Keep your tape recording to use in Exercise 5.)

   a. How does your voice sound? Monotone?
   b. Do you speak at an understandable rate?
   c. Do you use vocalized pauses (er, ahhh, umm)?
   d. Do you have a meaningful, vocal emphasis on important words?

4. Practice the speech you prepared for Exercise 3 in front of a mirror. Observe your facial expression, gestures, and bodily action.

   a. Does your facial expression change as you give different emphasis to various points in your speech?
   b. Are your hand gestures synchronized with your words? Do they add meaning to your oral expression?
   c. Do you move your head when emphasizing a point? Are you making eye contact with your imaginary audience? Do you occasionally move about, especially at transition points in your speech, to give visual relief to your audience and physical relief to yourself?

5. After you have practiced a speech before a mirror, as suggested in Exercise 4, give the speech in front of a mirror and at the same time record it on tape. Compare this tape with the tape made in Exercise 3.

   a. Do you notice any voice differences between the two tapes?
   b. Is there more variety in tone, rate, and volume in the second tape when compared to the first tape?
   c. Did your use of gestures and bodily action improve the effectiveness of your voice?

BIBLIOGRAPHY

Crocker, Lionel, *Public Speaking for College Students,* 3rd ed., American Book Company, New York, 1956.

This book stresses the importance of being able to speak well in public situations. Chapter 5, "Ways of Delivering the Speech," and Chapter 17, "Techniques of Structuring the Speech," contain many helpful suggestions and techniques.

Lewis, Thomas, R., and Ralph G. Nichols, *Speaking and Listening: A Guide to Effective Oral-Aural Communication,* Wm. C. Brown Company Publishers, Dubuque, Iowa, 1965.

This book attempts to coordinate the training processes needed for both effective speaking and listening. It includes exercises for practicing speaking skills and for evaluating both speaking and listening effectiveness.

Martin, Howard H., and C. William Colburn, *Communication and Consensus: An Introduction to Rhetorical Discourse,* Harcourt Brace Jovanovich, New York, 1972.

Chapter 4, "Understanding an Audience: The Listener's Alignment," contains an excellent discussion of audience analysis and assessment.

Ochs, Donovan J., and Anthony C. Winkler, *A Brief Introduction to Speech,* Harcourt, Brace, Jovanovich, New York, 1979.

Part 2 contains helpful chapters on outlining, speech construction, speech delivery, and use of the voice. The material is presented in a direct and practical manner.

Shrope, Wayne Austin, *Speaking and Listening: A Contemporary Approach,* 2d ed., Harcourt Brace Jovanovich, New York, 1979.

The book is aimed at helping one to develop the ability to communicate more effectively. A number of helpful tools are suggested for the speaker who uses visual aids and gives demonstrations.

9

Written Communications

Rudolf B. Schmerl

Writing about writing is dangerous business. It's not like telling people how to hit a ball, if you're Ted Williams, or build muscles, if you're Arnold Schwarzenegger, or have a flawless complexion, if you're Naomi Sims. Those people tell you about something they've been able to do, beyond argument, and they have the statistics or muscles or skin to prove it. But you can't tell people how to develop a prose style that will be universally admired. Not only doesn't the quality of prose style lend itself to objective verification, like a batting average; there would also be the implication that your *own* style, like Schwarzenegger's muscles, is overwhelming, or like Naomi Sims' skin, perfect. No, it's much easier and safer to criticize what others do or to write about what to avoid.

It may be that writing isn't just a skill. It *is* a skill, involving attentiveness, a sense of timing, techniques of one kind or another, and it does require practice to become better at it. But that is not all there is to it. "Language," an old Chinese proverb has it, "is the sound of the mind; and written language is a picture of the mind." Writing involves imagination and attitudes, not only of and toward the subject, but also of and toward the *audience*—the reader. What you write will be read under conditions you can only guess at. None of the clues that guide a conversation is available: your listener's nod of understanding, the encouraging smile, the quizzical eyebrow, perhaps the frown if you're going in the wrong direction. Nor can you choose the right moment for the reader's attention, the way you can often pick the time to tell your boss or your client about an unanticipated problem. You don't even have many typographical substitutes available for the normal audiovisual aids to conversation: you can't inflect your voice, you can't seat your reader in your most comfortable armchair and offer her a cup of coffee, you can't impress him with the cut of your clothes or the elegance of your gestures. All you can do is to draw a picture of your mind, of your mind *at work*. (There, by the way, in the italics, is one of those typographical substitutes—these parentheses, dashes, and commas are others—for audiovisual aids to conversation. The change of type face amounts to a signal to the reader that the writer is raising his voice. The parentheses suggest a self-interruption, accomplished, in talk, with a gesture as well as with a change in tone. The commas and dashes suggest pauses, spaces between notes as in Count Basie's piano.)

The kinds of writing discussed in this chapter—memos and letters, proposals, and reports—often rest on notes, sketches, and conversations summarized in some form, which cumulatively constitute a record of a process. That record can be important to a planner asked to explain why a particular project evolved as it did, taking unanticipated turns. The rationale has to be documented. But its documentation is not the same as its justification. Explaining it, and explaining it persuasively, are part of the planner's job.

"Communicate" is a verb that demands a direct object. Something has to be communicated. "It was great," someone will say about a "workshop" or a meeting, a conversation with relatives or Martians, "we really communicated with one another." The danger is not so much the lack of a direct object but the lack of awareness that it is missing. Once a speaker or a writer has lost his grip on what he is saying, he has lost his awareness not only of his audience but also of reality.

However tolerant we may be of such lapses in a speaker, we cannot extend a similar generosity to the writer. For all the advantages of immediacy the speaker has over the writer, the writer has an advantage of his own: an eraser. His work should be the product of reflection, revision, perhaps even of experimentation, insofar as he can show a draft to a discreet but honest friend and take her bilious reactions into account in his next version. As long as paper is cheap and waste baskets are large, pictures of the mind should be considerably more impressive than its sounds.

The writer's task is to combine a rhetorical technique—expository or persuasive—with a keen sense of audience and purpose. He has to be concerned with evidence for his assertions, the logic with which he develops them toward the conclusions he wants his audience to agree to, the clarity of his style (which is also the clarity of his thought), and the simplicity of his language. The chances are that his first draft will not achieve much of this, or not achieve it very well, and he should write it— whatever it is—accordingly: knowing that when he has finished it, he will have to go over it as if someone else had written it, someone whom he doesn't know. That is not news. But it remains unpleasant. Writing well is, in fact, often an exercise in self-criticism, which is so contrary to our national ethos, cultural dynamics, commercial orientation, and pop psychology that we invent one device after another to evade it: telephones, tape recorders, and dictaphones, editors who are called and paid to be secretaries, meetings so frequent and time-consuming that the production of even a first draft of anything becomes the production of the final draft as well, and most of all, committees, certain to blur not only language but also responsibilities. But the subject here is not how to evade writing. And to be fair, telephones, tape recorders, editor-secretaries, meetings, and committees are often all not only legitimate but even necessary parts of getting something written that exhibits a clear and consistent awareness of the audience.

Like most other professionals, planners write essentially three kinds of documents: memos and letters, proposals, and reports. It may be useful to discuss these as distinct kinds of written communications, although they may share purposes, audience, even rhetorical techniques.

MEMOS AND LETTERS

For the purposes of this chapter, a memorandum will be considered as a communication from one member of an organization to another: from a boss to a subordinate, from a subordinate to a boss, or from one colleague to another. A letter will be considered as communication from a member of the organization to an outsider. Distinctions between memos and letters having to do with forms of address, closure, and signature are largely matters of convention and—to give an example of the kind of conventional phrase found too often in both—need not detain us here.

The most important reason to write anything is that the available alternatives—to go see someone, to call a meeting, or to make a telephone call—are unsatisfactory in one way or another. Communication of anything takes place within a context of people, situations, events, and possibilities, whose characteristics and interrelationships sometimes seem clear to us when we decide what to do—write or call—and sometimes don't. Consider here the usual, specific reasons for writing memos and letters:

1. To ask some questions;
2. To answer some;
3. To provide information that hasn't been asked for;
4. To express a point of view that hasn't been asked for;
5. To announce and/or justify a decision;
6. To document previous communications;
7. To do some or all of the above.

Let's take these one at a time, a luxury rarely enjoyed by most professionals.

Aside from asking questions to exhibit interest or to stall for time, there are two things to think about in asking them: (1) Are they the "right" questions? That is, are they the questions that need attention now, whose answers will form a pattern for the actions to be taken to achieve the future that is desired? Or will they result in answers and activities of no direct relevance, of only secondary interest? Further, is the sequence of the questions logical in itself, and does that sequence of questions suggest that the corresponding sequence of answers (including the possibility that there are no answers!) holds the promise not only of illumination but also of agreement? (2) Are the questions put clearly and precisely? Can they be answered factually, succinctly, convincingly? How would *you* respond to them if you were their target? Asking questions raises more questions, and one you *don't* want raised is "Why do you want to know?"

Answering questions is usually easier than asking them properly. There are, first, the usual alternatives: "we don't keep our records that way"; "it would take more time than we have, with our limited staff, to compile the requested information"; "we hope to have the answers to your questions very soon, and when Mr. Newman returns from jury duty, will call your inquiry to his attention"—that sort of thing. Or, if your boss has sent you a list of questions demanding answers, you answer them: "In response to your questions: (1) Yes; (2) we estimate about 30 percent; (3) not more than three months; (4) $75,000; (5) we simply don't know yet." You might continue: "How do you want us to proceed?" But suppose the questions to be answered are not that straightforward. The boss might have asked you to summarize the issues and to make some recommendations. If the questions have to do with neighborhood development in Middlesville, you probably should write an outline of the issues, rather than an essay about them, grouped under headings (economic, legal, political, social; or immediate concerns, long-range problems; or some other division of topics and perspectives so as to impose some sense of order onto chaos, to facilitate comparisons, and to clarify trade-offs and alternatives). The recommendations would then rest on the description of the issues. Be clear about your intent: are you simply trying to explain something as best you can, or do you also want to persuade your boss of something?

Providing information that hasn't been requested calls, first of all, for a word or two about why you're providing it. Obviously, you feel that the information is important and that the recipient of your memo or letter should have it, but it isn't convincing to let it go at that. Most of us in large organizations are the almost daily recipients of in-house junk mail, marked "for your information." We all have more information than we need about matters of very little interest, and not nearly enough about the things we should be most concerned with. So explain why you are furnishing the information, without resorting to "I thought you ought to know that" or "I believe you will be interested to learn that," but in relation to some specific point. You may even want to *establish* the relation, as you see it. Then, after you've finished the memo or letter, ask yourself if it's really all that important, and what the likely consequences would be if you *didn't* send it. That, in fact, is not a bad question to ask of anything you write, say, or do.

The same is true about expressing unsolicited opinions. There are issues—personal, public, and professional—about which silence is or would be immoral and cowardly. There are others, however, about which it is or would be discreet and tactful. Unfortunately, at last report the human race has been unable to agree not only about which issues are which but also about the difference between cowardice and discretion, also known, since Sir John Falstaff, as the better part of valor. Would it be immoral or discreet to be silent about the newspaper's inaccurate and inflammatory attack on the planners' study of transportation modes in City Opportune? Again, what would be the likely consequences of speaking out, or the likely consequences of keeping still? They include having to live with yourself in either case.

Perhaps the most practical question the writer should consider— assuming he has decided, whether for visceral, moral, or calculated reasons, to write something—is how to express his unsolicited opinion in the most constructive way possible. He might, for instance, express his concern for truth, the future of the city, the newspaper's reputation and credibility. He might begin by denying the existence of any irreconcilable conflict, offering his point of view in the name of fact, sweet reason, and the public weal. And he could review his draft to excise any unnecessary adjective, any tinge of sarcasm, any trace of anger. Whatever his reasons for writing, the writing itself should not reflect them at the expense of focus on the issue. Of the questions he might welcome in response to his letter or memorandum, "Who asked *you?"* will not be among them. If any one kind of letter or memo calls for a critical reading by that honest friend before it is sent off, this is it.

Announcements of decisions of any consequence almost invariably include a description, however perfunctory, of the rationale behind them, or of the official rationale. Further, such announcements often include language intended to convey the sense that the decision has been made democratically, that is, that a majority of some kind has agreed to or endorsed it. Perhaps the governor of Hiatonka decides that he wants legislation offering large tax incentives to the builders and owners of such structures as hotels, office buildings, and factories that use solar energy collectors in their heating and cooling systems. The consultants' study of alternate energy sources provides the rationale. Nevertheless, the governor anticipates that the opposition will decry his proposal as "another tax loophole for the rich," and accuse him of having an eye on next year's senatorial contest. His writer is therefore likely, in drafting the announcement of the governor's decision, to refer to the consultants' study as "independent" and "thorough," to its recommendations as "compelling" and "farsighted," and to the proposal itself as "the next logical step taken by this administration toward the achievement of a sound and comprehensive energy policy for the benefit of all the people of this State."

The announcement of a decision often has less to do with the substance of what is being decided than with the conflict surrounding it. The more public the organization in which the decision is being made, the more careful seems to be the language used to announce the decision. The board of directors of a large corporation can fire its half-a-million-dollars-a-year president and appoint another without a great deal of explanation to the public. But a city administrator's decision to fire the $25,000-a-year director of the planning department had better be accompanied by some allusion to the community's future, in which a new spirit of cooperation and a heightened sensitivity to the concerns of all citizens will prevail.

When to Write and When to Call

The purely physical differences between air and paper, speech and writing, often influence the decision about whether to write anything at all. The Freedom of Information Act has resulted in bland, vague letters of recommendation about candidates for employment, often containing the suggestion that the writer would be glad to furnish additional information to the prospective employer of the person in question over the phone. The message is clear: the person in question is questionable, but the writer is too smart to say so directly. Letters can be copied; files can be examined; and it is harder to claim that what you have written has been misinterpreted than to assign responsibility to your listener for the meaning of what you have merely suggested in conversation.

On the other hand, you may decide to write rather than call precisely because your memo can be copied. Sometimes the *point* of a memo or letter is the "cc:" list after your signature, the carbon (now more likely the xerox) copies you send to the people whose names will stimulate some desirable action on the part of the official recipient of your letter. Or the strategy of sending a "blind" carbon copy to someone else, who is not named on the distribution list, can be a valuable defensive maneuver.

Both points may seem obvious enough. Nevertheless, any number of calls result in a request for "something for the file," and any number of memos and letters produce a flurry of telephone calls. No doubt much of this is inevitable, even necessary. But it is often worthwhile to consider not only what to write to whom, but whether to write anything at all.

On Dictation

The advocates of dictation, whether into a machine or to a stenographer, generally urge that the advantages include time and money (often regarded as equivalent anyway). Perhaps so. But the more important the communication, the more deserving it is of thought and reflection. Very few people can say very much very well about anything the first time. As a general rule, dictation should be reserved for the most routine kinds of written communications, the ones requiring almost no thought at all, which a secretary can cast into a pre-established format and perhaps even sign for you.

On the other hand, machines and stenographers can serve an enormously useful purpose for those of us still suffering from mental paralysis every time we confront a blank piece of paper. They can get us started. If you find yourself struggling to write what is clear in your mind, plug in the tape recorder or call the secretary. Once you can *see* what you've said, it should be easier to impose some coherence on it.

Writing Up, Down, Across, Out, and to the File

Writing to your boss, to a subordinate, to a colleague, to an outsider, and simply for the record requires some sensitivity to tone and nuance. Your personal acquaintance with the recipient of your memo or letter will, of course, be your best guide to what is appropriate and what is not. Still, it is often worth taking a moment or two to consider how you might feel about and react to your letter or memo were you in the recipient's position. The normal courtesies of language—"Dear Mayor"—become absurd if your first sentence calls the Mayor's integrity into question. The normal courtesies, in fact, are often insufficient, simply because they *are* only normal. Review your draft with an eye to its *possible*, not only desired, consequences. Is there a word or a phrase you might add, or should delete?

Writing to the file is similar to a letter following a conversation starting "This is to confirm our agreement reached in today's telephone conversation that" and so on. You want to get something into your records, factual, dated, for purposes perhaps still unclear, but just in case: often just in case an auditor shows up, next month or next year. This is the kind of letter you should be able to dictate without much trouble, because usually all you want in the record are facts, not nuances.

PROPOSALS

A proposal is a document intended to persuade the reader to approve a course of action described by the proposal writer. The approval is often to take the form of cash, called a *grant* or *contract*. In that sense, a proposal is intended to persuade the reader that his money would be well invested in the writer's purposes. It is an exercise in science fiction, describing how much better the world would be were the reader to give the writer the amount requested. It is an attempt to stick your hand in your reader's pocket *and* to make him feel good about it. Unless the proposal's success has been assured through prior negotiations, graft, or family ties, it is legitimately viewed as a sales pitch. A good proposal is a carefully worked out plan whose submission to the potential sponsor will include convincing evidence that everything required to carry the plan out is available except the money. That is what the sponsor is to supply.

The proposal business is extremely competitive. It is conducted in a variety of contexts simultaneously, one of which is the political context. Columbus persuaded his sponsors to finance his journey westward because they and he wanted to find a shorter route to the supposed riches of the Orient, and that kind of motivation remains discernible in government sponsorship of anything today. Given the understandable interest of our elected representatives in retaining their positions, the question of how to split whatever pie there is inevitably involves political answers. These have implication of critical importance to the planner and writer of proposals.

Planning the Proposal

Generally, proposals are part of a *process* in an organization, agency, firm, or institution, a process that is to result not only in a grant but in a well-administered grant, ending within budgetary limits and on time, and producing work almost as good as what was promised. The time to consider the problems the acquisition of the desired grant might produce—problems of space, for instance, or personnel, or even of money if matching funds are required—is when the proposal is being planned. Ultimately, good proposals rest on good ideas, carefully substantiated and planned in detail. That seems a safe enough truism. How can an organization try to create an environment in which such ideas are likely to flourish? There can't be a single answer, but perhaps the characteristics of such an environment can be described, at least as a set of ideal objectives.

First, there has to be an element of genuine dissatisfaction with the way things are: the central business district in Middlesville, or the transportation system serving City Opportune, or the present use of energy resources in Hiatonka and East Victoria. Proposals, after all, are about the future, which is to be somehow different from—better than!—the present. Just as dissatisfaction with ignorance is behind a good research proposal, so dissatisfaction with one or more aspects of a social system (town or city, state or region) lies behind a good planning proposal. That dissatisfaction needs to be expressed as concern for the future, of course, but it is dissatisfaction nonetheless.

Second, the organization has to know and understand the facts. Physical facts—square feet or miles, numbers of people and dwelling units, energy costs and retail volume—are absolutely essential. But by themselves, they are also insufficient. They need to be put into a variety of overlapping contexts: demographic, economic, geographic, historic, political, social, and psychological.

Third, the organization or its proposal writers should have ready access to relevant internal records, and should have some efficient, more or less routine, in-house way of communicating needs, opportunities, and procedures. *Vitae* should be kept in a standard format and up to date. Deadlines for proposals are announced as soon as they are known. Procedures are well publicized. People know to whom to turn for what.

Fourth, since proposal writing is a lot of hard work and since not everyone is equally proficient at it, proposal writers need to be adequately rewarded for their efforts. They should not be paid, treated, or made to feel like technicians, but like professionals as indispensable to the organization as any others.

Fifth, the organization is constantly aware of the political context in which it has to operate. That includes not only state and federal legislation, existing or impending, but also administrative rules and regulations, policies, priorities, and personnel of the agencies of interest. The organization knows not only legislators but also their aides, not only mayors and governors but also their executive assistants. It keeps in touch with, and its staff informed about, developments in City Hall, the state capitol, and Washington.

Sixth, the organization knows it takes money to get money. It has invested as much as it can afford in supporting services, including travel funds for proposal development, because it understands the importance of personal contacts. It is liberal about long-distance telephone calls, the use of the copying machine, and compensatory time off for secretaries who sometimes have to work on weekends to meet a deadline.

And finally, and perhaps most important, the organization knows that it is no better than its staff, and acts accordingly.

But of course organizations don't write proposals. People do. Let's say that you are looking at guidelines for a proposal that seem to call for a great deal of information, organized under major sections called "Needs Assessment" and "Goals and Objectives" and "Organizational Profile." There are also a number of formidable charts for the budget, which has to be presented in three or four ways, and a variety of forms demanding assurance of compliance with a host of laws, promulgations, regulations, and expectations. You know that the guidelines are the Bible—and you are absolutely right about that—but that doesn't help you get started. And you do have to get started.

The chances are that you will want, even need, the collaboration of other people. You probably know who most of them are. Begin by writing something to and for them: a rough draft or at least a detailed outline. And in a quite specific way, for a proposal—almost any proposal—has to answer at least six questions:

1. What do you want to do?
2. Why is it worth doing?
3. How do you want to do it?
4. How will you know how good it is?
5. Who is to do it?
6. How much will it cost?

The problem statement

The first question, often known more formally as "Goals and Objectives" is often the most difficult to answer. Technically, goals have come to mean your long-range or overall or ultimate aims, whereas objectives are supposed to be the more immediate mileposts you have to pass on your way toward the goals. But at this point, in writing something for your colleagues or associates, try to answer what you want to do as simply and clearly and especially as succinctly as you can. If you find yourself writing more than a couple of short paragraphs about it, stop. Go back to that section later and refine it. The longer the answer, the more confusing the reader will find it.

Significance of the problem

In the formal proposal, this section may be called "Needs Assessment" or "Background," but the basic question that needs answering is why what you want to do is worth doing. Here is your chance for explicit salesmanship. Implicit salesmanship should pervade the entire proposal, but here you demonstrate that you know your subject and you show how your idea is a step forward. Think like a lawyer: anticipate possible counterarguments, qualifications, questions. But a word of caution. Don't try to appeal to the reader's soft heart if you are talking about concern for the ill, aged, poor, and downtrodden. Don't, in fact, dwell on the misery in which your "target population," as the military metaphor has it, has been suffering for generations. You certainly need to provide statistical information that characterizes your population and its problems, data chosen so as to make your point and clarify your idea. But remember that your reader is likely to read other proposals, competing with yours, describing situations that might make yours sound tolerable. Don't compete in melodrama.

Procedures to be employed

This section will probably be the longest. To write it, you need to remember that the grant you want will be awarded for a specific period—usually twelve months, starting before the end of the federal fiscal year. You are to do whatever you're going to do in those twelve months, and at a cost not more than what you'll get. Assume you've gotten the grant. What will you do the first week? It had better not be recruitment of the key people; they should be ready to go to work immediately. What work? And when they have done it, what's next? What about the next weeks? What will have been accomplished by the end of the first three months? If you have to write a progress report to the sponsor after six months, what do you hope to be able to say? Some sponsors like to have the answers to these questions presented graphically—horizontal bars flowing from left to right, one beginning as another ends, representing progress as time passes—but at this point write it all out. You have to plan the work in detail, in an inherently logical sequence that will permit you to keep to the time schedule imposed by the award of the grant, and bar charts without sufficient narrative aren't much help to you or your reader.

Be sure to anticipate, in this section on procedures, what you intend to include in the budget. If some trips are going to be necessary, explain them in this section. If you have to lease space or equipment, indicate how much space or what equipment, why you need it—that is, why your organization can't provide it—and whatever else the reader should know about it. In some instances, a "Budget Justification" section is required by a sponsor as an accompaniment to the budget. You might use that to explain the arithmetic, if it is complex or obscure, but as a general rule the section on procedures should prepare the reader for what he will find in the budget.

Evaluation

This part might not need to be a separate section if you can build some sort of assessment into what you're doing ("formative evaluation"), assessment of the design and procedures as well as of the results. But it may well be useful at this point in your planning to consider the question of evaluation as a distinct matter. One of the people to whom your outline might be addressed—if you aren't an expert evaluator yourself—might be such an expert, with experience in the kind of work you're describing.

Personnel

Some sponsors are quite precise about the information they want about the personnel for the proposed project, and how they want it presented. Others let you hang yourself with too much rope of your own choosing. You should provide *vitae* for all key personnel, including consultants if these are to play a major role, and you should keep each of these *vitae* restricted to a single page. Include information about education (colleges attended, degrees earned, year of degree, major fields), positions held, professional experience that cannot be deduced from positions held, honors and awards, and publications and major presentations. The publications should have appeared in juried, reputable journals, should be related to the proposal's subject, and should be recent enough to suggest that their author is still breathing. This section is not to be taken lightly, and the preparation of *vitae*—which, by the way, are not to be confused with job resumes!—should not be left to the last minute. The reader is certain to be interested in the fit between the work to be done and the qualifications of the people who are to do it. All needed skills should be represented.

Budget

Requests for proposals—"RFP's"—often indicate the maximum amount available, usually including both direct and indirect costs. The immediate question is whether this amount is sufficient for the work to be done. If the answer is no—and it often is—the next question is whether the work to be done is described loosely enough in the RFP so that its scope can be modified sufficiently to permit its conduct within the available funds. If the answer is no, the question to be considered is whether *not* submitting a proposal at all would be a tactical mistake. Sometimes it may be wiser to submit a proposal quite unashamedly outside the stated budgetary limits or substantially different from the work called for than not to respond at all, simply because a submission that is rejected offers opportunities for follow-up, especially if you think that the sponsor's RFP is unrealistic. You will at least have had a chance to demonstrate your realism, have gotten your organization's name on the record, and have established a basis for future contacts.

But if these are not matters of concern to you, then it is questionable whether you should write a proposal for a contract or a grant you don't really want. It is sometimes advisable to try to get a sense of the competition, especially if you have reason to believe that the RFP is "wired"—written for a particular bidder. (That bidder may, in fact, have had much to do with the writing of the RFP.) Federal agencies occasionally hold bidders' conferences, at which (they promise) they will answer questions asked by potential bidders. Usually they answer the questions by reading from already published guidelines. The reason to attend these conference is to see who the competition is, and to assess your chances in that context.

Occasionally you may be asked to submit a preliminary proposal, indicating the amount you think you will need by rather broad budget categories instead of by line items. Sponsors usually screen such preliminary proposals to select those (a small number) to be developed into full, formal applications. Such opportunities should usually be accepted. They give you a chance to exhibit your interest, publicize your organization and its qualifications, and, if yours is among the preliminary proposals selected for further development, to get a head start on the formal application.

If you are working on an "unsolicited" proposal, whether to a governmental agency or to a private client, you need to have some idea of what the sponsor or client could afford to pay for the work you have in mind. Agencies all have annual budgets that are matters of public record, as are the grants and contracts they have awarded in the past. The ability of private clients to pay what you think the work will cost is probably more difficult to determine, but a few inquiries should provide some guidance. In either case, it is often a good idea to propose a plan of work that can be reduced or expanded (thus influencing the budget), and, furthermore, can be divided into phases. Begin with the essentials, and move gradually, not inexorably, toward the ideal.

You are sure to get help with or direction about the budget from people within your organization, but begin by presenting something that reflects your ideas. Set the budget up with one column more than all the sources of funds involved—your organization (if there is a matching-costs requirement), the sponsor, other possible sources (municipal and/or state), with the last or extra column reserved for the total. Put the time period ("From—To—") at the top of the page. If

BUDGET

From ______________ To ______________

	Agency	Sponsor	Other Sources	Total
PERSONNEL				
Project Director (H. Burwood), 50% time @ $35,000 p.a.	$ 7,500	$10,000	—	$17,500
Project Associate (C. Bamya), 100% time	—	15,000	—	15,000
Project Associate (R. Gardener), 100% time	—	14,500	—	14,500
Secretary (100% time)	4,500	4,500	—	9,000
Subtotal, Personnel	$12,000	$44,000	—	$56,000
FRINGE BENEFITS @ 15%	1,800	6,600	—	8,400
EQUIPMENT (Videotape recorder, cameras, projector, associated items)	1,000	—	$3,000	4,000
SUPPLIES	—	1,200	—	1,200
TRAVEL (Automobile only: 300 miles/month @ 25¢/mile)	—	900	—	900
COMMUNICATIONS (Telephone & postage) @ $50/month	—	600	—	600
DUPLICATION & PRINTING	—	2,400	—	2,400
MATERIALS (See notes)	—	1,800	—	1,800
CONSULTANTS, 5 days @ $250/day	—	1,250	—	1,250
COMPUTER TIME	—	1,500	—	1,500
TOTAL DIRECT COSTS	$14,800	$60,250	$3,000	$78,050
INDIRECT COSTS @ _____%				
TOTAL COSTS	$	$	$	$

Figure 9.1 Budget.

the time period coincides with your organization's fiscal year, you won't have to worry about a separate line, in the "Personnel" section of the budget, for salary increments. Otherwise you will have to prorate those increments for that fraction of the budget year in which they will have to be paid.

A budget is a plan, not a financial statement. Or more specifically, a budget is a financial translation of your plan of work. Its realism and honesty (you must use the real salaries of the people to be involved) have to be unquestionable. Figure 9.1 is a possible format for the draft you're preparing at this point. This budget reflects various assumptions that need to be spelled out here, if not in your draft to your colleagues. Say your agency or organization is a private company, in which case the line for "indirect costs" might read "overhead" or "administrative expenses." Say that your normal base for this calculation is salaries plus benefits, and that you usually employ a rate of 75 percent on that base. That would bring the total requested from the sponsor to $98,200. The trouble is that your request is not to exceed $100,000, and the winning proposal should probably come in with a bid well under that. So you may want to leave the indirect cost rate and calculation blank at this time, to see whether the boss wants to cut back the scope of work (for instance, by reducing the time fractions of the project associates), or reduce the normal rate in this instance, or both, or neither.

The guidelines specify a requirement for matching costs, which you have managed to meet very nicely by contributing portions of the salaries of the project director and secretary (and consequently, of their fringe benefits), as well as with the value of the equipment shown, some of which will, you are sure, be made available to you by the city. You are quite sure that no air travel will be needed (if it is, you need to specify it: destination, round-trip coach fare, number of people, number of days, a *per diem* figure, and an allowance for ground transportation). The consultants you have in mind are local people and therefore require only a fee, not reimbursement for travel expenses. (Note, incidentally, that consultants do not receive fringe benefits; from an accounting point of view, consultants are supplies.) The "materials" you have in mind for the $1,800 you've put in the budget might be kits, hand-outs, booklets, or some special aspect of your procedures you will have described in that section. Refer back to that explanation in your budget notes.

Supplementary materials

These will include the *vitae* of H. Burwood, C. Bamya, R. Gardener, and the local consultants; perhaps letters of support from strategically selected community groups or neighborhood associations; and possibly illustrations, graphs, figures, and tables, with which you did not wish to interrupt the text but which are important evidence substantiating your major arguments.

On Teamwork

Now it is time to talk to your colleagues. Send them what you've worked up after you've talked to them on the phone, making sure that they will be available for a meeting within the next two or three days. Attach a note reminding them of the date, place, and time. They should be asked not only for their reactions and ideas but also for whatever facts and figures they may have or can get that would fit into one or more sections of the proposal. If the basic idea is judged sound and you can work out the necessary modifications of the budget, you should discuss and agree on at least the following:

1. Is more information needed about the idea within your agency or firm? If so, who will get it, and when?
2. Is information needed from the prospective sponsor? If so, who will get it, and when?

3. Are there any implications, any potential conflicts, that should be explored? If so, by whom, how, and when?
4. If it is agreed that a formal application is to be developed, who is to do what and by what time?

Scheduling Production

The last question can be answered fairly readily if the people present at the meeting prepare a schedule, a kind of reverse calendar. The only firm dates are today's and the deadline's. Start with the deadline's date at the bottom of a piece of paper and then write the target dates, going up line by line, for mailing, printing, getting the right signatures, proofing, typing, writing the components of the final draft, and preparing the components of the rough draft—which should take you to the top of the page. Write the names of persons responsible, or agreeing to accept responsibility, for each component on the same piece of paper, and have copies distributed to all concerned as soon as possible. And now you should have some help in turning that outline into a fully developed proposal in accordance with the guidelines.

The Final Product

The more technical a proposal, the greater the need to assign the central role to the person most expert in the field. Again: a proposal is a *persuasive* document. It is not enough to describe something in general terms, nor even in specific terms, for description and persuasion are not the same. The sponsor's readers, more often than not, are sophisticated and knowledgeable people who need to be convinced that all parts of your proposal have merit and fit together. Someone other than the proposal's chief author, someone who is neither related nor indebted to him, should review the proposal for its clarity, logic, structure, tone, and persuasiveness.

You should be familiar with how proposals are routed or processed in your organization, how much time is required, and who is likely to be difficult about what. If you don't know, then find out before you embark on any of this. Something similar pertains to getting advice from the sponsor before you submit your proposal, assuming that you think it might be helpful. See if the sponsor would consent to review a draft of the proposal. It is doubtful that there will be much time for this, but if you have genuine technical questions, this might not be an unreasonable request. In any case, you should make some effort to get to know your sponsor's staff, especially if you hope or expect to continue to do business with them. They can be just as interested as you in putting a face and a handshake together with company stationery and a voice on the phone. And you are always likely to learn more in a face-to-face conversation than from those calculated memos and letters already discussed.

Advice can also be useful after the fact, i.e., after a turn-down. You can't expect to learn much on the phone, and it can take months to get the reviewer's (often guarded) comments in writing. But it's important to learn why your proposal was turned down as quickly as you can. After all, you're already at work on the next one.

REPORTS

Planning the Draft

Planning the draft of a report means thinking about three fundamental questions:

1. What is the essence of what I have to report?
2. To what use do I want this report to be put?
3. How can I try to get it put to that use?

A report is primarily expository. If it is also persuasive, it is not persuasive as salesmanship but as truth, or as close to the truth of the moment as you can get. A report on the conflict of case study 1, involving the proposed development of an urban neighborhood, might identify the various positions, the data supporting and negating each, their advocates, and the probable consequences of adopting, modifying, or rejecting them. A report about the experiences of cities comparable to City Opportune, which, in similar circumstances, had already made decisions about their transportation systems and are beginning to live with the results, might note the extent of comparability between these communities and City Opportune; might describe the factors leading to the decisions these cities had made; probably would describe those decisions in detail; and consider selected aspects of the accumulating consequences. The consultants' report about the possible uses of forms of energy other than petroleum in Hiatonka and East Victoria would presumably be laden with data about energy requirements, present and projected, by user category; about costs and benefits; about the ramifications of the various strategies considered, not only on the basis of experience elsewhere but also taking into account technological developments and political, economic, and demographic probabilities. All three reports would probably be lengthy and complex. But since they have very different subjects and very different audiences, their organization needs to reflect not some general textbook formula but rather the writer's careful consideration of the essence of what he has to say.

A general formula for the organization of the report might be (1) acknowledgments, (2) abstract or summary, (3) introduction, (4) scope or limitations, (5) methodology, (6) findings, (7) discussion, (8) conclusions and recommendations, and (9) appendixes. It is a prefabricated format with the usual advantages of prefabrication for both manufacturer and customer. It also has the usual disadvantages: it is unoriginal and undistinguished, it is tidier than thought or life, and it makes what is being said hard to remember, like a house looking like other houses on a street looking like other streets. If that is what you want, or what you have to do, all right. It is a way of organizing your material that won't get you into any difficulties. But if you have the time, freedom, and inclination to organize your material thematically rather than mechanically, you should consider the three questions posed at the beginning of this section. And to make decisions about thematic possibilities, you should think about the choices open not only to you but also to your reader, a subject to be discussed in a moment.

Collection, Analysis, and Compilation of Data

Three key questions about anything are (1) how much? (2) how many? and (3) compared to what? The last is by far the most important. It is the one that should guide what is to be collected, how it is to be analyzed, and how it is to be compiled or displayed.

In case study 1, for instance, the "how much" and "how many" questions are relatively straightforward: how much additional money is the proposed project likely to add to the city's tax revenues? In how much time? At what cost? How many people are likely to be affected directly? Indirectly? In what ways? How much time are the various options going to require for realization? But the third question is likely to demand a mix of *kinds* of answers, qualitative as well as quantitative, social as well as economic, psychological as well as physical, humanistic as well as technical. A merely statistical summary of the recorded history of the human race—births, deaths, geographical disbursement over time, average height and weight at selected stages of development, diet, diseases, reproductive patterns—would show how much of this and how many of that but, by neglecting comparisons, would probably leave the extraterrestrial reader ignorant of what he—or it— most needed to know: the human race is the most dangerous species on the Planet Earth.

That line of thought suggests an organizing principle for both the collection and the analysis of data. Begin with a tentative idea of how the data might be most usefully compiled, i.e., displayed so as to make important comparisons as easy as possible for the reader. Start, in other words, with some hypotheses about what you are likely to find as "raw data" and as the result of preliminary analyses; how, then, should you design the tables or graphs in which the data to be displayed will be most informative to the reader? No doubt you will revise the designs as you collect data you did not expect to find when you began, or as you find yourself in dead ends, or as analyses of what you've collected lead you in new directions. But begin with an idea of what you might want to show the reader, and how that might be done most effectively. It is almost always easier to revise and improve and add to a design than to start with nothing, accumulate a mass of information, and then try to puzzle out what it might mean to whom. It will also be difficult, under those circumstances, to meet the report's deadline.

Summaries

Abstracts or summaries are hard to write well, but they are likely to be the only parts of a report that a reader will look at twice, or perhaps at all. Further, they attract the attention of readers whom you may not have had in mind when you wrote the report itself: journalists, people with axes to grind, well-meaning but unsophisticated types who regard taxes as evidence of a communist conspiracy and poverty as a sign of the Lord's displeasure. An abstract should therefore be written for the benefit of a somewhat different, because wider, audience than the one to whom the report itself is addressed. It should capture the report's central point as clearly as possible; it should list (not describe!) those subsidiary matters simply too important to leave out; it should define, as precisely as possible, the report's limitations or scope; and it should be brief. You will probably have to rewrite it several times, pruning, sharpening, counting the words as you go. From 250 to 400 words should be enough for almost any summary.

Unless you have instructions to the contrary, the summary should follow the title page. Use some typographical substitutes for audiovisual aids, especially italics for key terms. A summary of the city planners' environmental impact statement in case study 2 might be as follows:

Summary

We have found *no convincing evidence* of any *probable deleterious impact* on the physical and social environment of City Opportune or its component neighborhoods of the proposed expansion of Route H23 by 3.25 miles from Rapp Road southeast to I-46. However, in view of certain concerns raised since the initiation of this study, we have included, *only for consideration, two alternatives* to the closing of Stuart Street called for by the original plan: (1) a covered, elevated *walkway* across H23 at Stuart; (2) an approximately 500-yard-long *underpass* of H23 to permit Stuart Street to remain open to vehicular traffic. Both are accompanied with *cost/benefit analyses* and *illustrations* of possibilities.

On Choices for Writers and Choices for Readers

Writing is obviously a matter of making choices, word by word, sentence by sentence. But so is reading. A reader can read everything attentively, or look for parts of a report to read attentively, or skim the whole thing, or pretty much ignore everything after the summary, just flipping the pages. The choices the

writer makes usually (unfortunately, not always) influence the reader's. Assuming that the writer is able to write clearly, even vividly, and that he has no problem with syntax, diction, grammar, logic, and the use of evidence, he may still want to consider the choices open to him in organizing his report as a whole. That is, in fact, the critical choice, the one that will determine all the remaining ones.

Involved here is a blend of two kinds of considerations: of the audience and of the material. It is something like giving a dinner party. If your primary consideration is the kind of food and drink you want to serve, you'll ask yourself whom to invite *after* you've decided that you want to display your recently acquired mastery of the preparation of delicacies popular in Trinidad. On the other hand, if the guest list is the independent variable, your selection of the menu and the sequence in which it should be served will presumably take what you know about the guests into account. The former possibility is analogous to the preparation of a proposal for which a sponsor remains to be found; the latter, to the preparation of a report for known readers expecting it, although unable to predict its contents.

Unless you are committed, for whatever reasons, to what has earlier been characterized as the formula approach to a report's organization, you should consider all reasonable and interesting choices you can think of for the way you might present the material. Write them down, as possible outlines. Some reasonable choices will also be dull, and some interesting ones will be unreasonable, but eventually you should arrive at a few that strike your fancy. The chances are that you will be able to characterize the key themes with which you might begin—and which will therefore establish the dominant perspective of, and set the tone for, the entire report—with an equivalent word: *money, equity, freedom, power, the future.* Now that you have completed the study, and are sitting in a pile of notes and computer printouts, what perspective makes the most sense to you? Would that be your answer if you were about to retire, or accept the offer of another job a thousand miles away? Does the whole thing add up only to a dilemma—on the one hand, on the other hand?

Write an outline of the report as you think it would be most effective, according to what you know or think about the audience, the subject, the content, and your purposes. Then put it aside and do something else for a while, preferably something entirely unrelated. When you come back to your desk, write another outline, different in some respects from the earlier one. Now compare the two. Is one clearly preferable to the other? Or are there aspects of both that you like and that could be combined?

The idea is quite simple: to make those choices as a writer that will influence the choices your reader has to make. Those are more predictable: to pay attention, to try to understand what you are saying, and to come to agree with your conclusions.

Supplementary Statements

Like proposals, reports often include appendixes—illustrations, tables of data, lists, letters, copies of legislation, maps, records of various kinds—intended to serve as evidence substantiating the main points made in the narrative. There are at least three things to bear in mind about these. The first is that they are often necessary. You don't want anyone to think that you aren't scholarly, and these appendixes, like bibliographies, references, and footnotes, are scholarly apparatus. The second is that you don't want gargantuan appendixes to make the report itself look like a pygmy. Be selective. If that's too difficult under the circumstances, see if you can present the report as volume 1 and the supplementary material as volume 2, which often simplifies matters for readers as well as writers. And the third is that whatever you include as supplementary material should have been anticipated in the text of the report, e.g.: "This point is thoroughly substantiated by the data presented in Appendix B, Volume 2."

ON REVISION: THE ROLE OF THE EDITOR

When you have finished the draft of your proposal or report, and before you take it to the boss or the copying machine, see if there is time for one more editorial review. You will have gone over it yourself, of course, and perhaps made some "final" changes, but the fact is that very few writers are able to separate their egos from their prose. If you have a professional editor on the staff of your organization, your draft will presumably wind up on his or her desk anyway. But if you don't, try to have someone only generally or even only faintly familiar with your subject go over what you've written—that discreet but honest friend.

A good editor is, in fact, everybody's friend: the writer's, the reader's, and the material's. A good editor's suggestions for revisions, whether stylistic or organizational, are intended for everyone's benefit. And no matter how pleased you may be with what you've done, presumably you will concede that it could be even better. After all, you aren't Ted Williams or Arnold Schwarzenegger or Naomi Sims.

A bad editor, of course, is another matter entirely. Luckily, bad editors are beyond the scope of this discussion.

EXERCISES

1. Start a file of clippings from newspapers and magazines about uses and abuses of English. Add notes of your own about what you hear on radio and television, and in conversations and presentations.
2. Write a memo about the need for more information about the issues involved in the development of the urban neighborhood (case study 1). Address the memo to your boss, who seems to you too anxious to press ahead. Write another version of the memo to a colleague whom you don't particularly like, trust, or respect, but who is influential in your organization. Write another version to a subordinate of yours about whose competence you have some doubts.
3. Dictate a letter to a client explaining why your report about some aspect of case study 1 will be at least three weeks late. Then ask a friend to criticize it and compare his or her criticisms with your own revisions.
4. Outline a proposal to conduct a new study of alternatives for the development of the transportation system of City Opportune. The proposal is to be submitted to the U.S. Department of Transportation. Now outline a proposal for a similar study, but of smaller scale, to be submitted to the city.
5. Outline a report to be submitted to an interstate committee appointed by the governors of Hiatonka and East Victoria. The report's findings are that planning to meet the two states' projected energy requirements, including anticipated demographic and economic changes, and taking possible legislation as well as existing laws into account, will remain inadequate without long-

term major involvement of the federal government. You are aware that this conclusion will not be popular with the committee.

6. Comment on the following memorandum:

STATE UNIVERSITY OF HIATONKA

Department of Urban and Regional Planning

MEMORANDUM

Date: July 15, 1981
To: Department Faculty
From: F. T. Priestley, Chairman
Subject: Graduate Directorship

As you know, A. Bernard Phillips will be Acting Associate Dean this coming academic year. After completion of that responsibility, he will be returning to his regular teaching duties and not to the Graduate Directorship. I regret this decision of his, but I understand it and am grateful for the outstanding job he has done.

To find a new Graduate Director, I convened a joint meeting of the Advisory Committee and the Graduate Committee. Due to summer schedules, three people were unavailable but the rest were able to attend. Subsequently, with the unanimous consent of this group, I asked Lawrence Lorch if he would accept the Graduate Directorship. I am very pleased to report that he has agreed to assume the office, and accordingly, I have appointed him Graduate Director, effective September 1, 1981. I am confident that he will justify the strong support he has received.

BIBLIOGRAPHY

Proposal Writing

Hall, Mary, *Developing Skills in Proposal Writing*, 2d ed., Continuing Education Publications, Portland, Oregon, 1977.

Provides a full discussion of proposal writing including many useful examples of federal forms and checklists. Highly recommended.

Lauffer, Armand, *Grantsmanship*, Sage Publications, Beverly Hills, 1977.

Fund raising and proposal writing in human services programs. A good, brief readable discussion of fundamentals. Includes examples.

Report Writing

Damerst, William A., *Clear Technical Reports*, Harcourt Brace Jovanovich, New York, 1972.

Recommended particularly for its emphasis on readers. Separate chapters consider letters, memos and informal reports, formal reports, and proposals. Exercises.

Mathes, J. C., and Dwight W. Stevenson, *Designing Technical Reports*, Bobbs-Merrill Co., Indianapolis, 1976.

The subtitle of this book is "Writing for Audiences in Organizations." Unusually well laid out and designed. Highly recommended.

Nelson, J. Raleigh, *Writing the Technical Report*, McGraw-Hill, New York, 1952.

A standard textbook that is particularly well illustrated with examples.

Trelease, Sam F., *How to Write Scientific and Technical Papers*, M.I.T. Press, Cambridge, 1958.

Recommended particularly for its discussions of outlining and organizing the material.

Style

Baker, Sheridan, *The Practical Stylist*, Thomas Y. Crowell Co., New York, 1962.

Although "primarily for freshmen English," this is an excellent book "useful . . . to anyone who must face a blank page and the problems of exposition."

A Manual of Style, 12th ed., University of Chicago Press, Chicago, 1969.

A standard reference work. Three sections ("Bookmaking," "Style," and "Production and Printing") include chapters on rights and permissions, punctuation, illustrations, captions, and legends, tables, citing public documents, and design and typography. Produced by the editorial staff of the University of Chicago Press.

Strunk, William, Jr., and E. B. White, *The Elements of Style*, Macmillan, New York, 1959.

One of the most famous and delightful discussions of style since its publication in the 1930s.

Reference

Hodges, John C., and Mary E. Whitten, *Harbrace College Handbook*, 7th ed., Harcourt Brace Jovanovich, New York, 1972.

One of many good standard references on grammar, punctuation, spelling and diction, effective sentences, paragraphs, outlines and organization, taking notes, and business letters.

10

Graphic Communications

Hemalata C. Dandekar

Planners who are also architects, landscape architects, or engineers tend to draw. They lavish affection on graphic presentations as a primary and almost exclusive communication medium, while planners trained in the social sciences do not draw and often use graphics only when they cannot avoid them. Either position leaves considerable room for improvement in effective communication. Although it is not true that every picture is worth a thousand words, graphics of various kinds are often essential for clear communication in the planning profession. There are verbal ideas and there are nonverbal ideas, and each requires the right medium to convey the message. The right graphic in the right place, at the right time, can be instrumental in making a good decision, winning a project, or clinching an argument.

The use of graphics in planning is constrained by the limits a planner imposes on his visualization of the problem. The extent to which graphics are used in projects should and does vary greatly depending on the type of planning project at hand. Generally physical planning projects necessitate more use of graphics and less written material than social, economic, or policy planning projects. In addition, however, it appears that those planners who know how to draw easily and well will often have great difficulty writing a three-page memo but "talk" (sketch) with pencil and tracing paper in hand while others, who do not draw, will omit using even the simplest maps in places where their inclusion would be very helpful. This state of affairs is unfortunate. With ever-evolving graphic techniques, often mechanized, it is important for planners to become familiar with the range of graphics available for use and to learn to integrate them throughout the planning process. A good practitioner does not have to be able to draw a desired graphic product himself but does have to know where a particular kind of graphic is needed and be able to explain clearly what is needed to someone who can execute the work. In short, the ability to conceptualize applications rather than to know specific techniques is most important.

Practitioners who allow the lack of graphic skills to

prevent them from using illustrations in their various professional communications are denying themselves the use of an evocative communications technique. In conjunction with written descriptions and statistical information, often expressed in graphic form by charts, diagrams, and graphs, graphics form the backbone of communication in planning. In the layperson's mind, in fact, maps, charts, and diagrams are probably the major tools of the planning profession.

Learning the intricacies of a particular graphic form is a long-term endeavor involving highly specific, technical, and professional skills. Schools of architecture, landscape architecture, and graphic design require many hours of class work to teach the skills of their trade. The studio setting, providing experience and opportunities for personal demonstrations, is the best place to learn these techniques. This chapter provides an overview and conceptual guide to where and how graphics can be used in various parts of the planning process, and why.

Some of the more conventional materials, tools, and technology of various graphic media are briefly described. The types of graphics used in planning and the considerations that go into the selection of quality, technology, and cost are discussed. Some of the obvious but often overlooked conventions of graphic presentation are listed, concluding with a brief review of various do's and don'ts of graphic organization and presentation.

The objective is to stimulate planners to use graphics effectively. Ever-more sophisticated technology enables even the most untutored in drawing and drafting skills to use graphics. However, any suggestion that future developments will make discussions of graphics anachronistic—that sophisticated technology and computers will suffice—is rejected. Commonsense organization and thinking through of the use of visuals in the planning process, as stressed here, do not become dated. People have used symbols to convey concepts fundamental to their existence for too long to shed the habit. A machine may perhaps execute these symbols and graphics in technically more finished form, but the imagination and sense of which ones to use where and how to use them effectively comes with discretion and the cultivation of a professional judgment for which there is no substitute.

Further, planners also work in countries where the supply of electric power is not dependable, and the means of transporting equipment are often incompatible with delicate machinery. In addition, in the most developed of countries too, sophisticated equipment can and does fail at the most critical moments. It is judicious, therefore, to remain committed to the simplest and least expensive techniques that will do an effective job of communicating.

The ability to convey ideas with the most basic tools—a stick to draw images in the sand, chalk and a chalkboard, pencil and tracing paper, relying not primarily on drafting skills but on the ability to conceptualize and communicate with universal graphic symbols—is a skill that will not be rendered obsolete by high technology. Communicating directly with graphics opens up new ways of understanding spatial relationships for both the individual making the drawing and the observer.

USES OF GRAPHICS IN PLANNING

Graphics are used at various stages of a planning project for various purposes. They range from simple diagrams and sketches, useful in thinking through a problem, in the preliminary schematics stage to complex illustrations used in oral presentation, written reports, and in legal, contractual documents. The type of graphic medium used in any planning project is determined not only by the scale of the project, its budget, and time frame but also by the projected role and perceived relative importance of graphics among the various means of communication.

A major consideration is to determine the audience. Who are the decision makers? What is the objective of the particular task for which you are considering the use of graphics? Some audiences are more receptive to graphic communications in the form of slides, maps, or drawings than others. It may be essential to use certain types of graphics to reach some audiences whereas others would find them too simplistic. Deciding what is appropriate must be carefully considered in the light of one's expected audience. Graphics are used in planning toward various ends. Some of the major uses, in order of increasing complexity, follow.

1. To illustrate existing conditions at a place. A picture can serve to authenticate oral or written descriptions. For example, the developer's team in Middlesville might take photographs of the existing dilapidated structures on the proposed site to corroborate their verbal description of the dismal conditions existing in the area at a city planning meeting. In turn, the tenants' union might take a picture of the Middlesville skyline under particularly flattering light conditions to illustrate how aesthetically pleasing the area is and how destructive the proposed high-rise development would be. Sketches, line drawings, and annotations on maps can serve the same function; for example, the map in Figure 10.1 illustrates the pressures on a proposed site of expansion of the surrounding land uses.

2. To record and document in a snapshot fashion an existing set of relationships: spatial volume relationships of built form, natural resources (aerial photographs of the extent of forest areas in Hiatonka and East Victoria), or human and animal activity and use of spaces and facilities (Junior Planner's photographs of people in Liberty Plaza).

3. To analyze relationships over time by comparing photographic or movie film and videotape recordings or other visual documentation such as artists' sketches or paintings, or diagrams and maps made over time. One such use was mentioned in Chapter 1. Time-series photographs of people's activities in a popular urban space in New York City were taken by planners to analyze how people used that space at different times of the day through the year. Another use, familiar to agricultural economists and geographers, is the mapping of landownership over a number of years to see if there are significant patterns of change. (See Figure 10.2.)

Another, a technique widely used in planning, is computer-generated maps of census tracts that show the geographic distribution of a particular variable. Such techniques provide different spatial pictures of a phenomenon, pictures that can contribute toward better conceptualization and therefore, one hopes, to better analyses. Excellent examples of this are maps generated for the widespread metropolitan area of Los Angeles, which consists administratively of numerous cities and several counties. A table for a variable derived from data obtained in the census tracts in the basin, such as the frequency of single female heads of households with dependents in each tract, can never have the same immediate and vivid impression as a computer-generated map. The map can be designed to show the higher frequency of occurrence of tracts in darker tones. The poorer downtown areas of Los Angeles and the corridor down to Watts show up dramatically and convincingly in such a map as dark, highest-density areas. (See Figure 10.3.) The visual image, far more evocatively than a statistical tabulation, shows the area's disparities.

4. To be evocative and elicit an emotional response. Pictures and films of baby seals being clubbed are

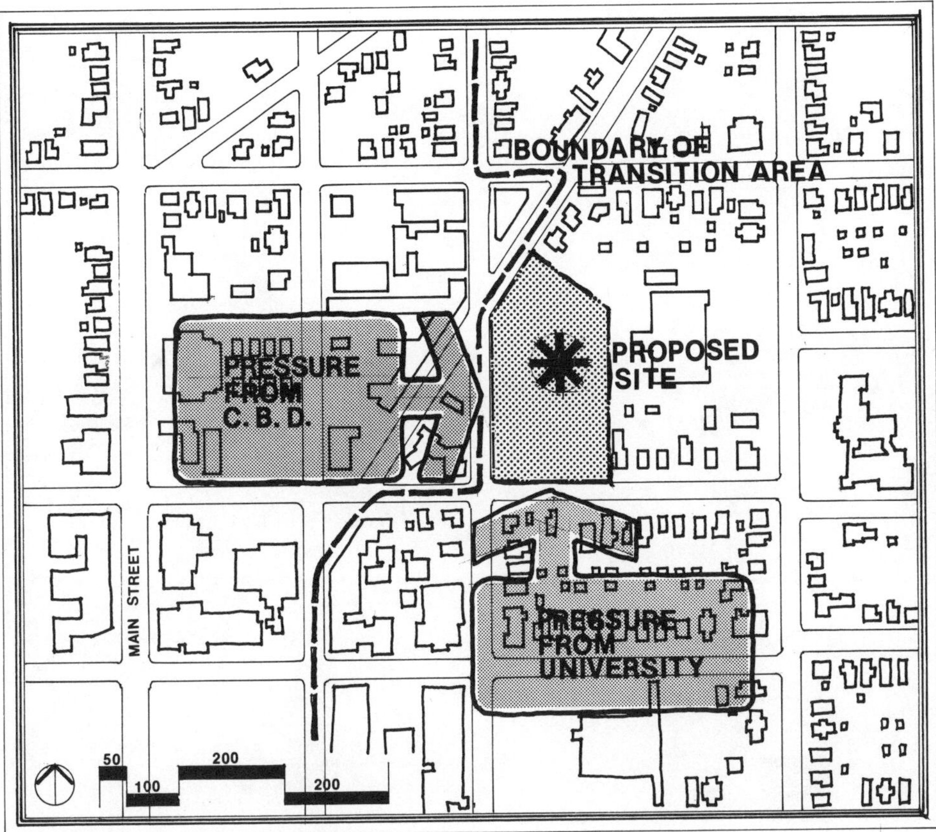

Figure 10.1 Land use pressures on proposed site.

well-known instances of such use by conservationists to get support for legislation to ban this activity on the Northwest coast. In one presentation made by a neighborhood group fighting to protect the existing use of a neighborhood urban space from urban renewal encroachment, the strongest argument was a slide of an old man sitting on a bench, eyes closed, blissfully sunning his large and exposed stomach. What better use of a public open space for utilization by the poor could one have? Old photographs and illustrations that show how things used to be can also often be evocative if skillfully used in a presentation to reinforce a point of view.

5. To communicate new ideas and interventions. Graphics can be used to show how things might be. For example, the Middlesville developer's planning team could juxtapose a slide of the existing dilapidated condition of the adjoining street frontages with a slide

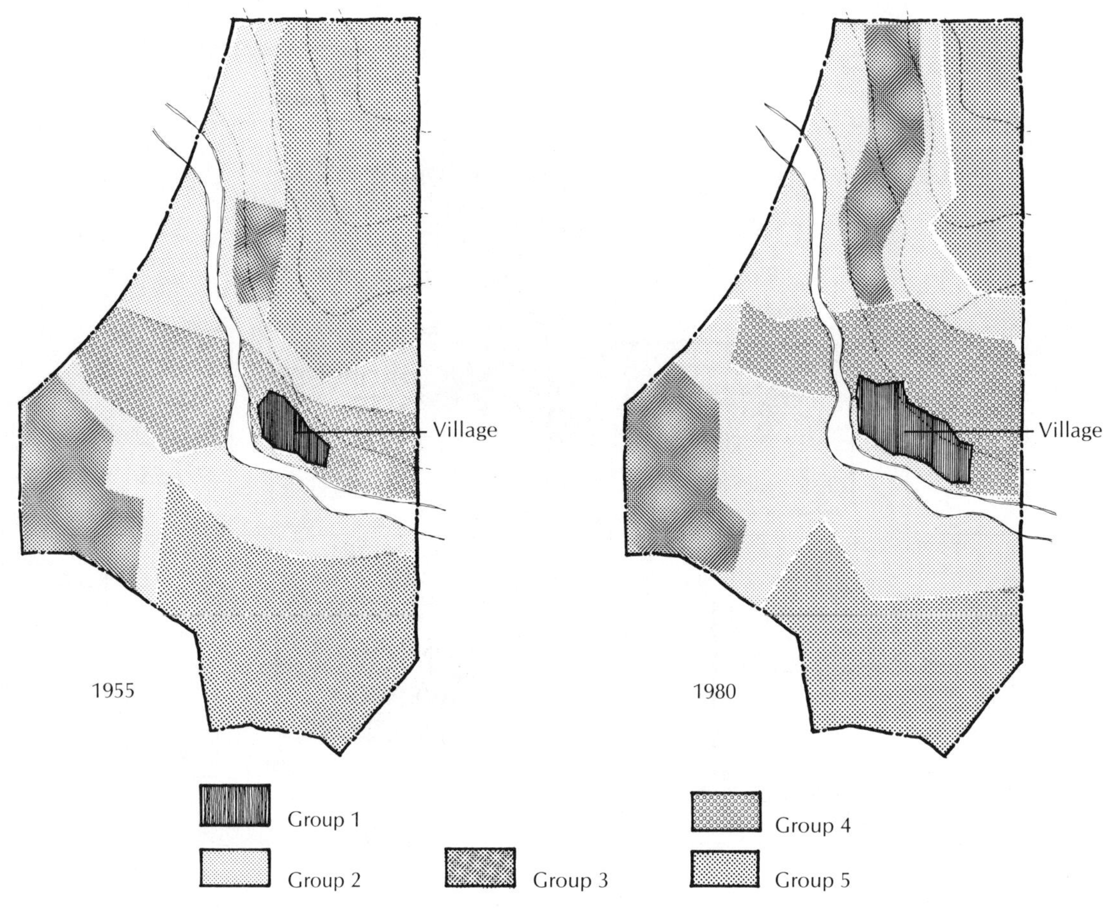

Figure 10.2 Ownership of land in 1955 and 1980.

showing, in overlay, the same scene with their proposed project drawn in highly flattering terms so that it appears to improve the area. (See Figure 10.4.) Before-and-after pictures of an intervention are not the sole prerogative of product commercials.

Details of master plans for physical changes in existing parks or streets can be sketched in order to communicate a general approach to the redesign. For example, Figure 10.5 could be used to communicate to citizens in Middlesville how a currently low-use park in a neighborhood might be improved so that many more user groups would benefit.

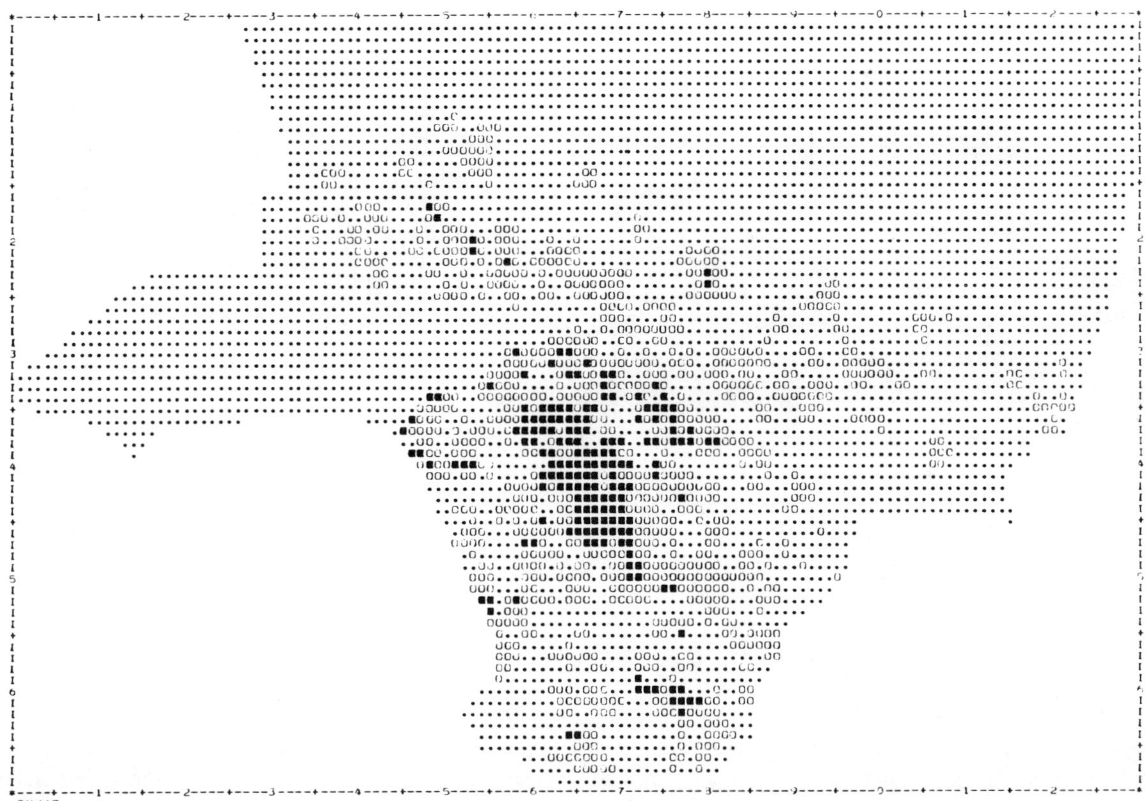

Figure 10.3 Single female heads of households in Los Angeles. (Map produced by the School of Architecture and Urban Planning, University of California, Los Angeles, using the SYMAP program supplied by the Harvard Computer Graphics Lab.)

When and Where Used

Graphic material is used throughout the evolution of a planning project from preliminary site reconnaissance and review of secondary sources to the final presentation of findings to various groups and the compilation of the report and other contract documents. Different types of graphics are generally used in the different phases of a project. A typical project might begin with maps, photographs, sketches, and overlays to document the site conditions. Flowcharts, diagrams, and organizational charts might

Existing

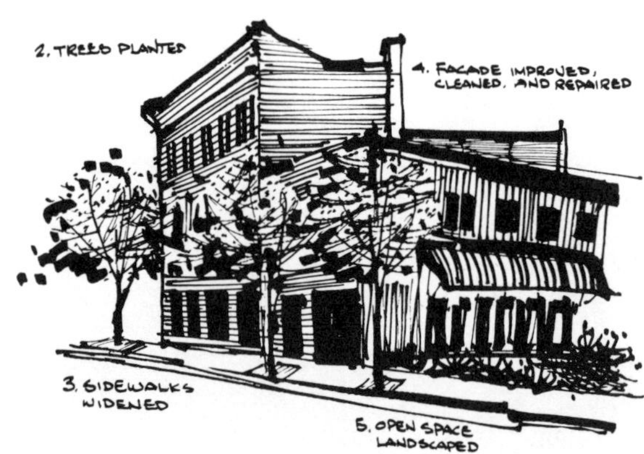

Proposed

Figure 10.4 Proposed renovation of streetscape.

be added to reinforce proposals. Concepts might be developed using bubble diagrams, which provide an understanding of relationships and interactions, and matrixes could be constructed to establish priorities between alternatives and build group consensus around decisions. Descriptive material in simple form might be produced for release to the mass media. Much of this material might be molded and changed to be compiled as part of the final report.

Some of the ways graphics are used in various planning stages can be illustrated by describing what the developer's team of planners in Middlesville might put together to convince enough groups that the proposed development is good for the city at large.

Proposals

The planning team may show on a base map of the site and adjoining properties the various amenities and annoyances currently there. They may reinforce this with photographs and slides. Along with these, they may produce a series of overlays on the base map to illustrate alternative proposals for development, with the annoyances removed and amenities preserved. Giving these positive and negative values, matrixes may be drawn, as illustrated in Figure 10.6, to show, quantitatively, which alternatives, including maintaining the site in its present use, might be the most beneficial. Obviously such comparisons are valid only if people agree with the assigned values. As part of a series of items of evidence, they will make an impression. A trial lawyer knows that the judge may tell the jury to disregard some presented evidence that is inadmissible, but once it is heard, the impression remains.

Proposals often contain flowcharts and diagrams to explain how the proposed work will be completed, the time frame, the actors involved, and their work relationships. The strength of their interaction and the increases or decreases in their participation in various phases can be indicated by different line widths and weights. One way to do this is as illustrated in Figure 10.7. Proposals should integrate what has been learned from secondary sources, from site visits and observations, and the conclusions arrived at for a program of analysis and intervention based on the

Figure 10.5
Park development.

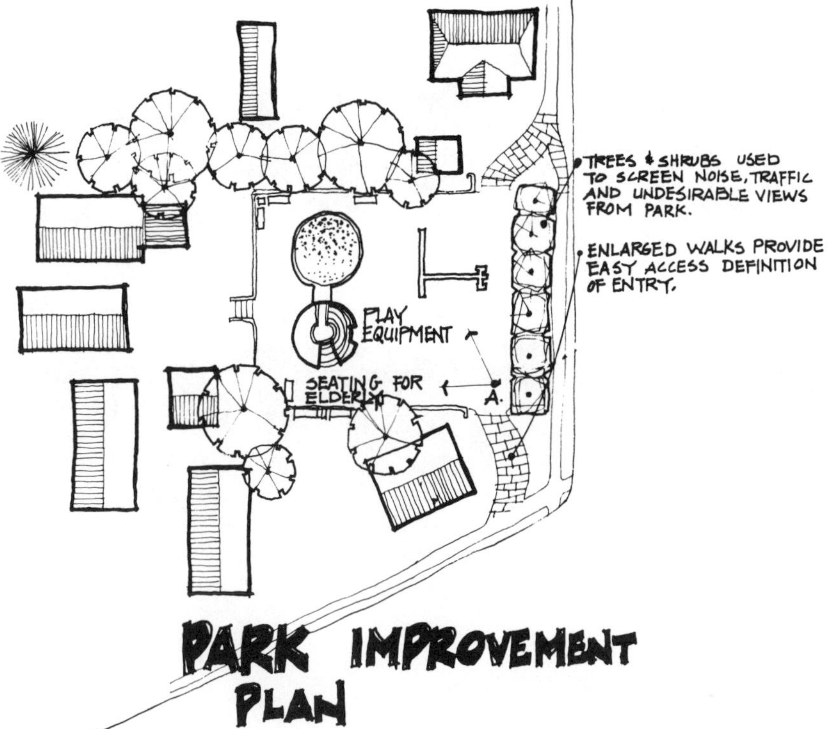
TREES & SHRUBS USED TO SCREEN NOISE, TRAFFIC AND UNDESIRABLE VIEWS FROM PARK.
ENLARGED WALKS PROVIDE EASY ACCESS DEFINITION OF ENTRY.
PLAY EQUIPMENT
SEATING FOR ELDERLY
A.
PARK IMPROVEMENT PLAN

View of improved park from point A

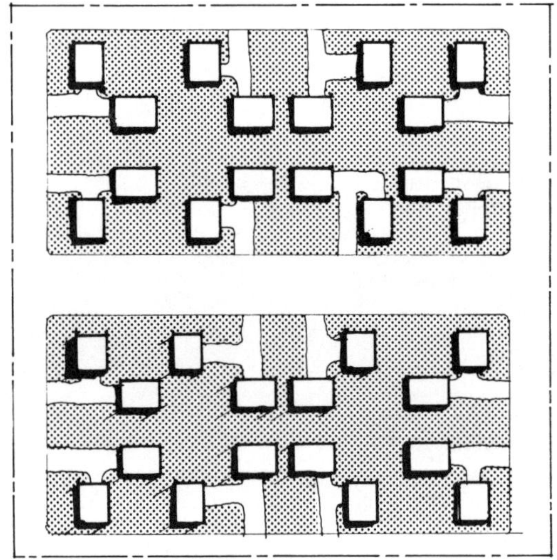

Existing single detached housing

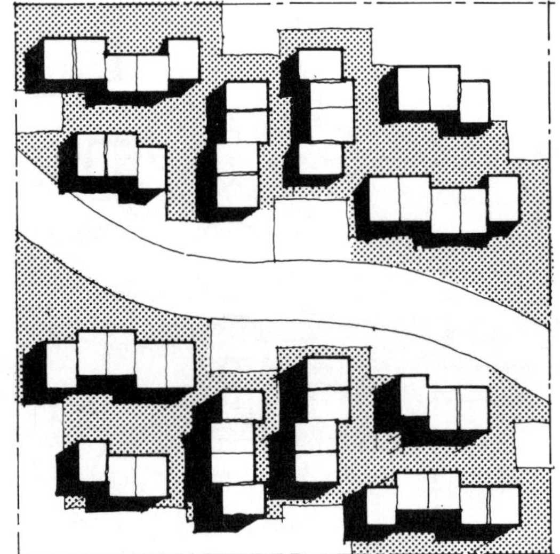

*Alternative 1: medium density townhouses

Alternative 2: high density high-rise apartment

Evaluation Matrix

Attributes / Alternatives	Existing low density housing	Alternative 1: medium density	Alternative 2: high density
Neighborhood interaction	-1	4	1
Neighborhood indentity	1	3	2
Land utilization	-1	2	4
Access to services	2	2	1
Total	3	11 *	8

*Alternative 1: medium density townhouse is the more desirable alternative.

Figure 10.6 Evaluating alternatives.

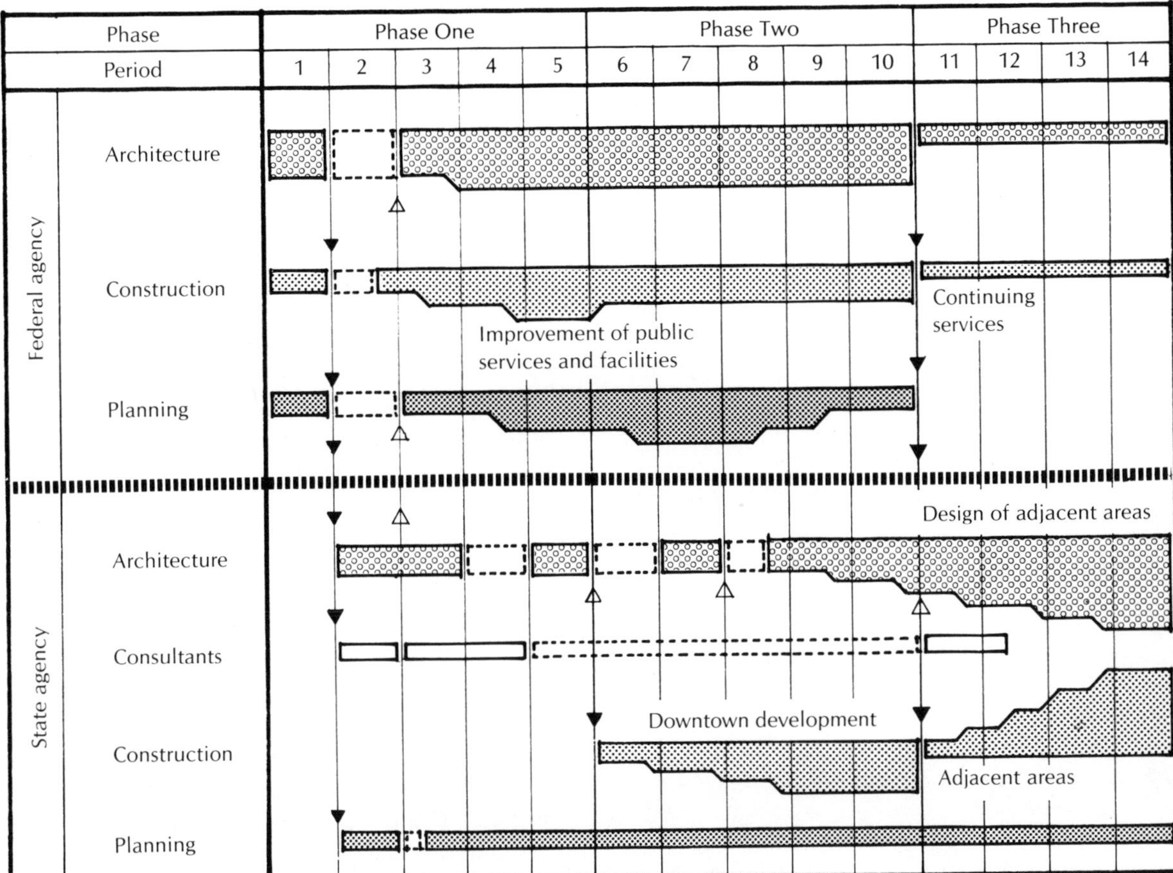

Figure 10.7 Flow chart.

preliminary review of the material. Flowcharts can help reflect this thinking in a tangible, spatial way that provides a time frame.

Site visits

Site visits may be carried out to update maps and to photograph and document site vegetation, topography, available social services, condition of housing, or proximity to shopping and schools. It would be useful, for example, for the developer's team in Middlesville to illustrate on a map how the proposed development might improve local business by bringing people to within walking distance of the downtown commercial establishments. Maps can be used to show connections, travel time, and traffic flows, and juxtaposed and keyed with photographs to indicate what a walk in the area might feel like.

Small group discussions

Bubble diagrams, maps and overlays, illustrations of products, plans, and sketches on tracing paper that are rough and conceptual in nature may be used to communicate and develop ideas during group discussions of a project either in-house or with consultants or clients.

Presentations

Graphics are created for presenting projects to clients in small or large groups, in formal and informal settings. They may consist of presentation and delivery of legal, final documents or informal, intermediate documents delineating a phase of the project in roughs.

Memos, meeting reports, and recommendations

These may contain an assimilation of various graphics used in earlier phases and specially prepared illustrative maps, key maps, key sketches, reproduction of photographs, and similar material for the specific document.

Final reports and recommendations

This is one of the most important uses of graphics since the final report or document containing suggested recommendations goes on file and is referred to whenever questions pertaining to the project come up. It is the legal document that fulfills some of the planner's contractual obligations and will tend to be widely disseminated to various constituents.

The graphics and other visual material used in these reports must be clear as well as accurate. Maps should have correct scales and give sources of the data displayed. As far as possible, the graphics should be restricted to black and white, or to colors that can easily be reproduced in black and white, to reduce the cost of reproduction. Watercolor washes and felt pen colors look attractive but are expensive to duplicate. Even though color reproduction is now readily available, it is expensive. In addition, reproduction, particularly by photocopiers, can change the color in a variety of unexpected ways. If color is necessary, test runs of the colors that are expected to be used on the copy machine to be used will show how each color reproduces. Small areas of a color can show up unexpectedly dark and dense in a reproduction.

Mass media releases

Usually for cost reasons the visual material provided for the media must be black-and-white reproducible originals. Clarity and simplicity of images, whether photographs, maps, or other illustrations, are the primary criteria. Only the two or three major things that must be conveyed should be included; more will clutter the image.

TYPES OF GRAPHICS USED IN PLANNING

Tables

Cross-tabulations and simple frequency counts are the most commonly used statistical measures in planning. Displayed in tabular form, they allow the reader to look at the numbers in more detail than was described in the text and to examine other, perhaps related parameters of the problem. They substantiate what is said elsewhere in writing or verbally. When tables indicate three or four categories (see, for example, Figure 6.3 where low, medium, and high levels of participation are illustrated), using graphic symbols, as has been done by use of full, half, or empty circles in this table instead of letters or numbers, can make for a more quickly comprehensible image.

Bubble Diagrams

These are used usually to clarify and conceptualize the nature of relationships and interactions between individuals, groups, or spaces in the system under

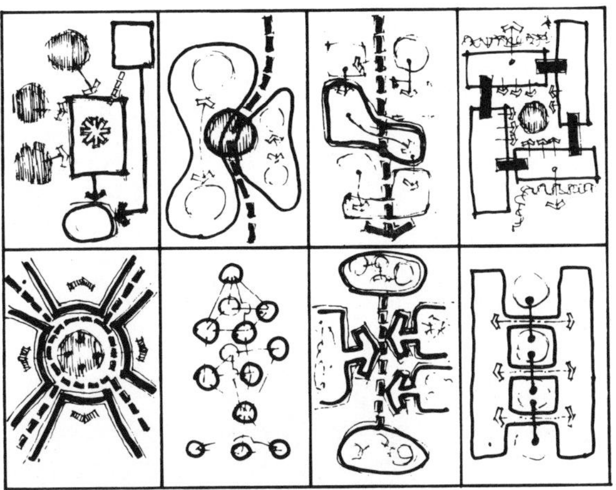

Figure 10.8 Bubble diagram.

observation. For example, a bubble diagram can be overlaid on an existing map showing neighborhoods of a city to define the boundaries of various ethnic group settlements and to indicate the flows and connections between them. Such a diagram can be used to identify and cluster land uses (at a macroscale) or spaces in a building (at a microscale).

A bubble diagram can also be used to conceptualize the groupings and connections of various individuals in an organization. The connecting arrows and lines and shading in overlapping areas can define the interactions and relationships, whether hierarchical or parallel, as illustrated in Figure 10.8.

Bar Charts, Graphs, and Diagrams

These are used extensively in planning to provide a succinct and quickly comprehensible overview of data and statistical information. Bar charts such as histograms are useful in giving a picture of relationships at a point in time or visual comparisons of changes over a given period. Graphs can summarize a series of observations. They are helpful in indicating current trends and can be used in making future projections. Diagrams can be instrumental in simplifying and conceptualizing complex relationships in graphic form.

Flowcharts and Organizational Charts

Flowcharts are often used to indicate long- and short-term scheduling and programming for a planning project. Flowcharts can be designed to communicate vividly the parallel or discrete activities involved in a particular project, the skills, resources, and the needed time frame. They provide a tangible image of the time period in which defined activities must take place. Insofar as planning is related to programming and organization of actions in the future, flowcharts are immediate evidence of a planner's contribution.

Organizational charts are depictions of networking and hierarchical relationships between groups and individuals in an organization. They are very useful in conveying the formal power structure and relationships. They can also be constructed to show the informal system of connections. They are useful in the process of analyzing important connections and bottlenecks and for developing strategic interventions and reorganizations. (See Figure 10.9.)

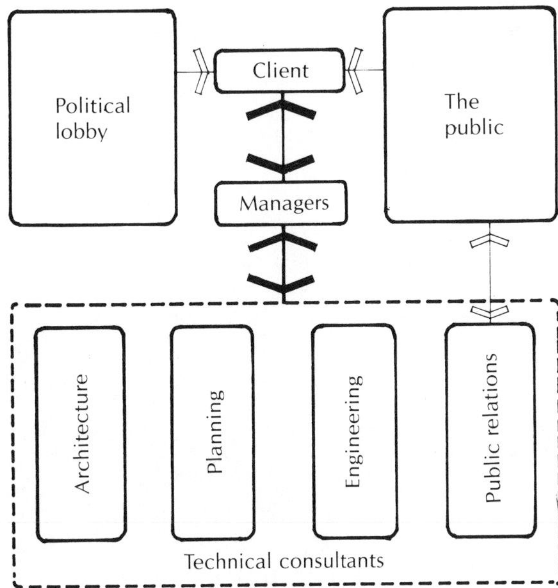

Figure 10.9 Organizational chart.

Maps

Maps used in the planning profession are of various sizes and subjects and differ in scale, complexity, and detail. They include site and neighborhood plans, survey maps, topographical maps, census block and tract maps, transportation and land use maps, utilities maps, and aerial photographs, and range from a local city scale to vast areas of natural resources such as forests, taken by satellite photography. Infrared photographs are used for land use planning or detection of pollution in air and water, energy consumption, and much more. Some uses of maps are to locate industrial sites, plan reclamation and conservation strategies, do land-use analyses, plan energy development, monitor strip-mining, define flood-prone areas, plan solid waste storage, and design continental shelf exploration.

In the presentation of a project, a series of maps of increasingly larger scale can draw attention to a particular site or area in a larger context. Such maps locate the project in a larger regional, national, or even global frame. Figure 10.10 is a set of maps showing the country and state; the pertinent characteristics of the immediate region around Middlesville that are conducive to future growth of the town; and the project site. The set of maps is used to reinforce the developer's position that additional housing is needed in the area. Such maps are good heuristic devices that can help widen the viewer's perception of the problem and have him or her consider it in a larger, geographic-spatial-issues context.

In these maps the designer must be careful to show only those elements he wishes to draw attention to. A map is drawn to point out and stress a particular set of relationships. The key, therefore, is to simplify, to curb the instinct to provide an overabundance of information, and show only what is important. Such conceptual maps should provide just enough information about significant areas or landmarks and names of adjoining and arterial streets so that the observer is oriented. But too little as well as too much information can be misleading. It is important, even

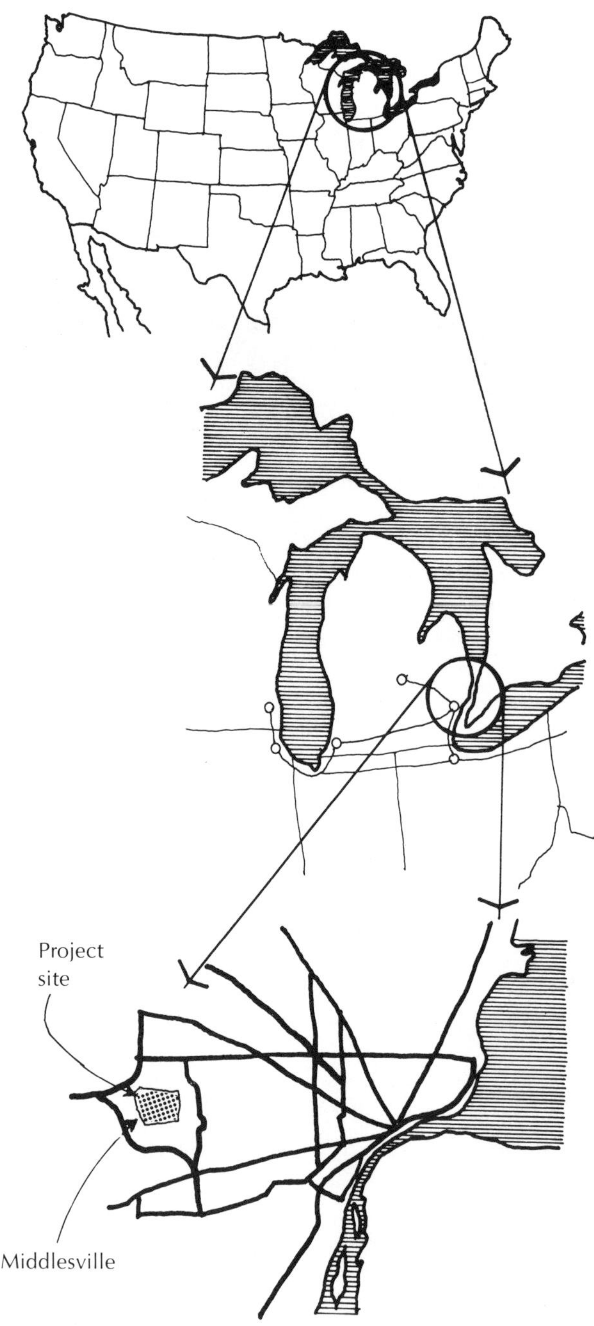

Figure 10.10 Maps to establish context.

in schematic maps, to give some idea of scale, to keep north facing the top of the page, to give sources of information, and to have the map as updated as possible.

Maps in conjunction with drawings can be used to "walk" a viewer through a project. In such cases the maps and visual material should be laid out so that ready reference from one to the other is possible. Maps can show more than just the location of objects. Human activities take place in space and so can be defined, observed, located, and translated into maps that show their relationships. Variables like socioeconomic conditions and demographic traits can be depicted to show their spatial distribution. In conjunction with symbols such as arrows, dots, and other indicators, maps can reflect the dynamics of a particular situation. For example, you can use dark, thick, and light, thin or dotted lines for heavy and light traffic flows, dotted and solid arrows between defined areas to indicate pedestrian movements, and so on. Color overlays on polyester film or tracing paper can be used to show overlapping activities.

Drawings

The types of drawings used in planning can range from quick freehand sketches rapidly made on tracing paper, newsprint, or a chalkboard during a discussion or presentation to finished, drafted, and accurate drawings that can become legal contractual documents (for example, the master plans for a large recreation complex or a new town development). Computer drawings are becoming increasingly prevalent, especially in uses requiring repeated renderings of base information.

Cartoons and Caricatures

Cartoons and caricatures can be particularly powerful if developed with the right blend of professional insight and humor. Cartoons often have been used to communicate expeditiously issues of social policy or power relationships in a proposed planning project. The tenants' group in Middlesville might use cartoons depicting the eviction of local residents by rich developers in order to mobilize the neighborhood quickly.

Three-Dimensional Models

These are widely used by physical planners at scales ranging from studies of parts of a building to urban design and city master plans. Gaming and simulation techniques often make use of boards and various artifacts to simulate land, regions, resources, buildings, and services. In gaming the opportunity for participants to see, modeled in three dimensions, what planning processes at work do to the physical plant leads to a high level of involvement. The participant can get a visceral understanding of the systems at work.

Photographs and Slides

Black and white and color photographs and slides are widely used in planning documents and presentations. With ever-evolving improvement in the design of cameras, and a simplification of their operation, almost anyone can take successful photographs and slides. Although professional photographers are hired for high-budget projects requiring sophisticated work, most planners will be required to do their own photography on a number of occasions and must become familiar with the art of taking usable slides and photographs. Good photographs that are well integrated into a report or slides that are properly sequenced and thematically interjected into an oral presentation are relatively inexpensive and a powerful means of communicating information.

Videotapes and Movie Films

Video and movies are used to document existing conditions or planning processes at work. For example, visual recording of citizen meetings, construction of infrastructural projects like dams, and delivery of social services such as health care can be vividly descriptive. Video and film are also used to record events, to disseminate information about particular case studies, to trigger discussion, and to educate. The costs of producing films are relatively high, but the impact of a well-made documentary can also be commensurately high. Videotapes are cheaper to produce than films and are an excellent way to present three-dimensional information in presentations to which models cannot easily be shipped without damage.

GRAPHIC TOOLS

A large range of tools and materials is used in creating graphics for planning projects. The simplest and possibly one of the more versatile is a drawing pencil. Pencils come in various types, from very soft, almost charcoal-like consistency to very hard. Hard pencils are numbered H, 1H, 2H, 3H, up to 9H in increasing order of hardness, and soft are numbered B, 1B, 2B, up to 6B in similarly increasing softness. Also there are leads labeled F and HB in increasing softness, between an H and a B. A line weight varies according to the pressure applied by pressing down on the lead in drawing. The same drawing lead feels different on various paper types and finishes (the more tooth or grain a paper has, the harder the lead need be to get uniform thickness) and the drawing surface (harder surfaces make leads feel softer). The factor that cannot be controlled is the humidity. High humidity tends to make the leads appear harder. Thus it is necessary to experiment with various lead weights to see which will provide the correct effect. As drawing conditions change, so will the pencil lead. Generally 4H to 9H are used for work requiring a high degree of accuracy, 3H to B for architectural line drawings, lettering, arrows and scales, and 2B to 6B for freehand work, rendering, and details. The work normally done in an average planning firm can easily be done with leads ranging from 2B to 2H.

There was a time when first year drafting techniques instructors in architecture would come around, break a laboriously sharpened pencil lead, and ask the student to sharpen the pencil with a pencil knife. They did not believe in mechanical pencil sharpeners and other gadgets, and students spent precious time whittling away at their stubs. Fortunately, teachers no longer make such archaic demands. Commercial establishments have developed easy methods for obtaining the required thickness of pencil leads. There are a variety of mechanical pencils that consist of a lead holder, ready leads, and some easy tools to sharpen them. The standard lead holders allow one to change leads rapidly for different softnesses. The microlead pencil holders are the latest and most useful devices since the leads rarely need sharpening; they are a specified thickness ranging from 0.3 mm to 0.7 mm.

When you are using a straight edge, such as a T square or parallel rule, and want to make a consistently thick line, you must rotate the pencil in your fingers as you draw. Otherwise on a very long line, the lead and the line get fatter toward the end. The microleads are somewhat less prone to this problem.

Ink pens of various types and felt-tipped pens are other commonly used materials. Technical drawing pens that use waterproof drawing ink to give a dense, dark line are made by various manufacturers and are available in a range of thicknesses. They make drawings look finished since they can give, by virtue of the consistency of the pen point, a uniform line. The way to get a uniform-thickness line width is to draw the line slowly, holding the instrument close to a vertical position, so the ink has a chance to flow in the necessary thickness. Stopping in the middle of a desired line will tend to result in a slightly enlarged ink blob there. If you do stop, pick up the pen and start with a tiny gap from where you stopped. It is a good idea to do this even when working with pencils or felt pens. Ink pens come with instructions about how to keep them clean. It can take hours to clean a point, particularly one of the thinner ones, if the ink has dried in it. Some pens are claimed to be self-cleaning, but my experience is that none is perfect and all must be constantly maintained so they do not fail, particularly during the last touches on a final presentation.

Felt-tip pens also come in various thicknesses and are available in colors. It is impossible to get uniformly solid consistent colors with felt pens. Every time you stop a stroke, you will get a darker deposit. This problem is less acute with some papers and more acute with others, so quick tests must be made before working on the final drawing. If you have small areas to color in, you can go across the full area in one or parallel strokes, and then you have only to accept the slightly darker areas of overlap between parallel lines as part of the overall pattern. You can regularize the overlaps, both horizontal and vertical, and accentuate them to give a texture. Make sure your supply of felt-tipped markers is replenished and fresh when you start producing final graphics for a project. In addition, it is wise to check for consistency between old and new markers in the lighter tones such as beige, sand, light olive, and gray. The same name on the label does not always assure that the colors will be identical.

Black felt-tip pens are useful for making bold, simple sketches, for bubble diagrams, and for creating simplified, stylized maps. The latest development in felt-tip pens is micropoints that come in accurately gauged sizes. Since felt-tip pens are much more convenient to carry and take care of, they are preferred on many occasions to ink pens. The microtip felt pens also offer accuracy and can be used in applications where formerly ink pens were mandatory. They do not, however, give as dense a line and therefore do not give as clearly defined a line on prints. (Figure 10.11 shows a variety of drawing tools.)

Areas can also be colored with pencils or pastels. The former do not usually give the vibrant colors of felt pens but produce a softer, more delicate effect. They are used to color drawings on various kinds of boards or drawing papers. Skillfully handled, pastels can be used on yellow or white tracing paper to produce quick, quite delightful drawings. They are bold, and used flat, on edge, they give broad, dramatic lines. Thus they are a good medium to use on newsprint or yellow tracing in front of an audience. Pastels can also be used to add exquisite touches of color and highlight to black-and-white line drawings. Practice and a few demonstrations from someone knowledgeable can be quite helpful in learning to use pastels skillfully. Pastel-colored drawings photograph well into color slides, the pastels showing up as vibrant colors. Cut-outs of solid colored papers stuck

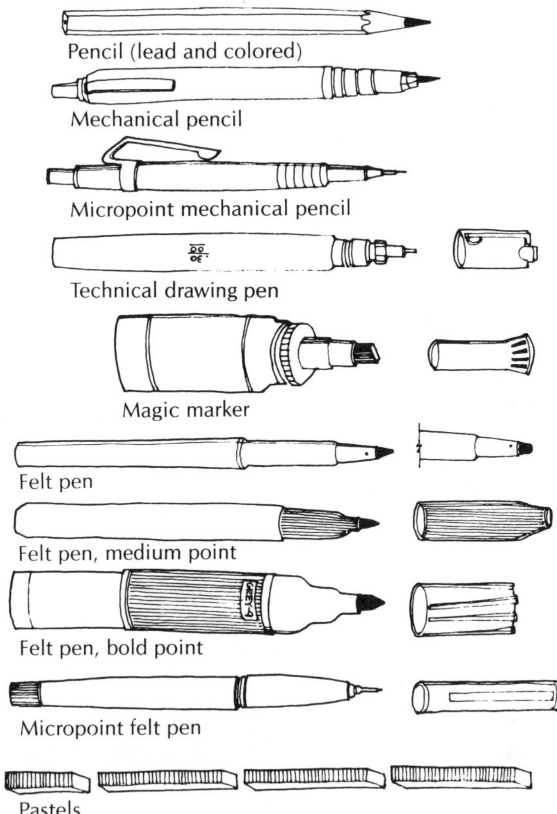

Figure 10.11 Drawing tools.

onto background paper with or without writing is another simple, bold, and effective, as well as quick and easy, technique.

There is a large array of papers to draw on. One of the most useful and versatile of these is the range of tracing papers available. Tracing paper comes in a variety of weights, which is tied to the quality and the cost of the paper. Heavier weights are sturdier and are used for drawings that might be reproduced many times or kept as a permanent record. They are also more expensive. The flimsiest, and traditionally the most popular among architects for rough conceptual sketches, is yellow tracing paper, fondly known as canary paper or onionskin. Available in standard widths, starting at twelve inches and going to forty-two-inch rolls, this paper is relatively inexpensive. Being transparent, it is used to develop basic ideas on successive overlays. Like canary paper but slightly heavier in weight are rolls of white tracing papers available in the same widths and used basically in the same way. More precise tasks are usually done on the white tracing paper because it is a little sturdier.

Tracing paper can be laid on maps. Only the pertinent information from the map is traced. This tracing can be used to make multiple blue, black, brown, or sepia line prints on papers that have been treated with a diazo emulsion and that can be colored or added to in various ways. Sepias are diazo prints made on tracing paper, which can in turn be worked on and then used as originals to run off additional prints.

More permanent drawings such as base maps, basic building outlines (sketch elevations), and frontages may be made on superior quality tracing paper. This is of two types: medium grade (weight 16 lb.) with a fine or medium tooth, which is used for preliminaries, and quality grade, generically called vellum (16 or 20 lb., 100 percent rag) used for finished drawings, both of which are available in twenty-four-inch and thirty-six-inch-wide rolls. Polyester film 0.004 mils. in thickness is also commonly used for drawings and, next to linen, is the most durable and dimensionally stable and gives the cleanest print quality. It is available in clear sheets, which are used to make overlays on maps and charts. Polyester film is treated so as to be rough grained and capable of receiving pencil and ink lines on one side (which is thinner and cheaper) or both sides (which is double weight and more expensive). Gridded film has grid lines that help in maintaining a module throughout a drawing but do not show up in print. Film is available in standard-sized sheets from 9 X 12 inches to 24 X 36 inches or in thirty-six-inch-wide rolls. Special leads must be used because normal ones will cut through the film. And ink pens used on it must have special tungsten-carbide steel points, or the pen point will soon be rendered useless.

Most architectural and planning firms use polyester film for their working drawings and other more permanent graphics. Base information, such as the boundaries of the site and existing vegetation, or the street patterns of a city, can be printed on the reverse or back of a polyester film or sepia diazo print, and can be drawn and erased on the front without affecting the base information. When base information is to be printed on the back of a reproducible print, a reverse-reading sepia or polyester film is needed. Sepia reproducible prints are less expensive than film and they are also less sturdy, but if the work is short term or tentative and conceptual, then they are a moderately priced good choice.

Illustration boards are used either for finished drawings or for mounting prints for presentation. One hundred percent rag boards, medium weight or heavier, are suggested. Models are often constructed out of Strathmore illustration boards, which are dense and white clear through. Rough models are also made from boards that have a core of plastic foam with thin cardboard on either side. These, available under various names such as Foam-core or Art-core, are about a quarter inch thick and available in sheets 20 X 30 inches to 32 X 40 inches.

Various devices make lettering, shading, and

drawing conventional symbols easier. They include different sizes and types of plastic stencils for lettering, various kinds of templates for drawing circles and to indicate different plumbing fixtures, and on-site equipment. Rub-on lettering is available for legends and labels, as are rub-on trees, cars, and people and other often-used symbols available at various scales. Well-articulated hand lettering and sketching is quicker and cheaper for those who are skilled, but it is not essential to become proficient at it. But often one can just trace over the outlines of an image on a tracing sheet. In hand lettering it can be helpful to use a small triangle to serve as a guide to get straight verticals. Various kinds of lettering machines are available but they are time-consuming to use and generally give a mechanical-looking lettering devoid of character.

Tone paper of various densities and textures is available for shading. You must be careful to burnish rub-on lettering and tone paper carefully and firmly. When printing, the original is put on top of the light-sensitive paper that the print will appear on and passed under the lights in the print machine. The areas not exposed to the light where you drew in pencil or felts do not get developed and so appear as dark lines or areas in the print. The emulsion side of the printing paper is thus facing the back of the original, and the front of the original, facing the light source, gets heated. The heat or the lights tend to burn off the rub-on lettering, which can get dislodged from the original and lodged onto the rollers of the printing machine. Some rub-on material has special adhesive and is designed for originals that will be run through the printing machine. Even so, great care must be taken in applying it and checking that it remains well adhered.

It is very demoralizing to find the original art work, on which you have spent many laborious hours, in a sorry state of disrepair, rub-on letters askew or missing. If there are only a few, restricted areas of lettering, these can be covered with clear sticky tape to protect them; however, in some prints the tape will show up as a halo around the lettering, which is not desirable in fine-quality work. A piece of clear acetate placed over the whole drawing while printing can solve the problem of burn-off. Another way to avoid it is by making the first print a reverse-reading sepia or polyester film and using that for subsequent prints. Reverse readings are made by placing the original facing the emulsion side of the printing paper and its reverse side facing the light. In this case the rub-on lettering is not so likely to burn off. Stickers with the project name, name of your firm, or other identification can also be made that have adhesive fronts and can be attached on the back of the polyester films or tracing, thus providing consistent labels from sheet to sheet. Conventionally this block of information is placed either along the bottom edge of the sheet or the right edge. North arrows, scales, date of printing, date of revisions, and similar information are often all arranged within this block. (See Figure 10.12.) Various weights and finishes of white drawing paper are available for more conventional drawings, sketches, and watercolors. Currently, however, most graphics in planning are made on transparent sheets because they allow for easy duplication.

Other equipment is helpful: French curves to help in drawing or tracing over segments of curves, small and large triangles that are 45 or 30/60 degrees, adjustable triangles that can open from 45 degrees to anything up to 90 degrees, T-squares (which are less expensive and getting to be somewhat old-fashioned), and the more expensive but simpler to use parallel rule. To get parallel lines the T-square has to be held firmly in position against the edge of the drawing board by the pressure of the nondrafting hand; a difficult task for the neophyte draftsman. Much simpler to use is the parallel rule, which enables one to make parallel lines easily because it moves up and down the drawing board on a system of wires and pulleys. The more expensive models of the parallel rule have the advantage that they are kept off the paper surface by the roller bearings they slide on. It is therefore much easier to keep drawings clean and smear free. A T-square can quickly turn the beginner's efforts into

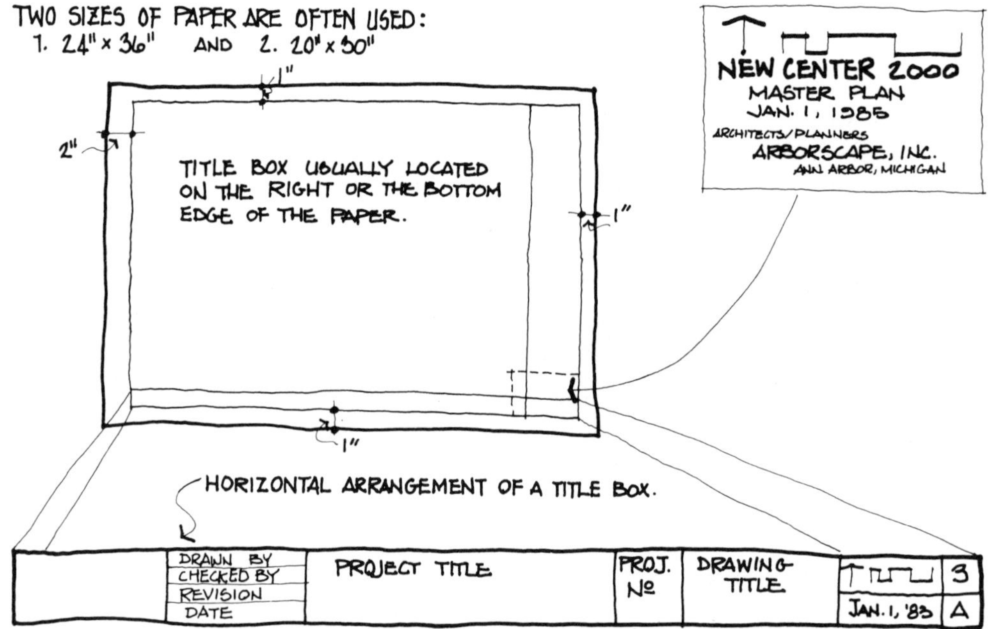

TWO SIZES OF PAPER ARE OFTEN USED:
1. 24" x 36" AND 2. 20" x 30"
1"
2"
TITLE BOX USUALLY LOCATED ON THE RIGHT OR THE BOTTOM EDGE OF THE PAPER.
1"
1"
NEW CENTER 2000
MASTER PLAN
JAN. 1, 1985
ARCHITECTS/PLANNERS
ARBORSCAPE, INC.
ANN ARBOR, MICHIGAN
HORIZONTAL ARRANGEMENT OF A TITLE BOX.
DRAWN BY
CHECKED BY
REVISION
DATE
PROJECT TITLE
PROJ. Nº
DRAWING TITLE
3
JAN. 1, '83
A
TITLE BOX ALSO CAN BE PLACED VERTICALLY. IT INCLUDES BASIC INFORMATION SUCH AS PROJECT TITLE, NAME OF FIRM, LOCATION, PROJECT NUMBER, DRAWING TITLE, SHEET NUMBER, NORTH ARROW, SCALE, DRAWN BY WHOM, AND DATE OF DRAWING. FOR PRELIMINARY DRAWINGS, TITLE BOX CAN BE MUCH SIMPLER AS INDICATED IN THE EXAMPLE IN THE UPPER RIGHT CORNER.

Figure 10.12 Typical sheet layout.

grey smears. If that happens and if you have pressed hard on the lines you want, exposing the print for a longer time to the light while printing could give a usable copy. Other special items are rubber-covered scales that can be bent to help make projections on graphed curves or to draw over irregularly shaped curves as for site boundaries and roads. (Figure 10.13 shows some of these tools.)

Photographs, publicity materials, fabrics, and maps can be combined into collages that can be reproduced in black-and-white format on a copying machine. They can be designed to convey an ambience of the place. Duplication machines of many kinds are versatile tools for planners. They can make reductions, enhance contrasts, and often produce prints that look much better than the originals.

Simple line drawings can be easily constructed from slides by projecting the slide on to a drawing surface, laying tracing paper over the image, and outlining. The problem is, as you draw, your hand is likely to interfere with the projected image. To get around this in a visuals studio, you may find, or be able to set up for yourself, a glass box so that the image is projected from the back of the glass, and you can trace over it by placing tracing paper over the image. (See Figure 10.14.) You can thus make a copy of the outline without interfering with its projection on to the screen. You can also make a tracing from a

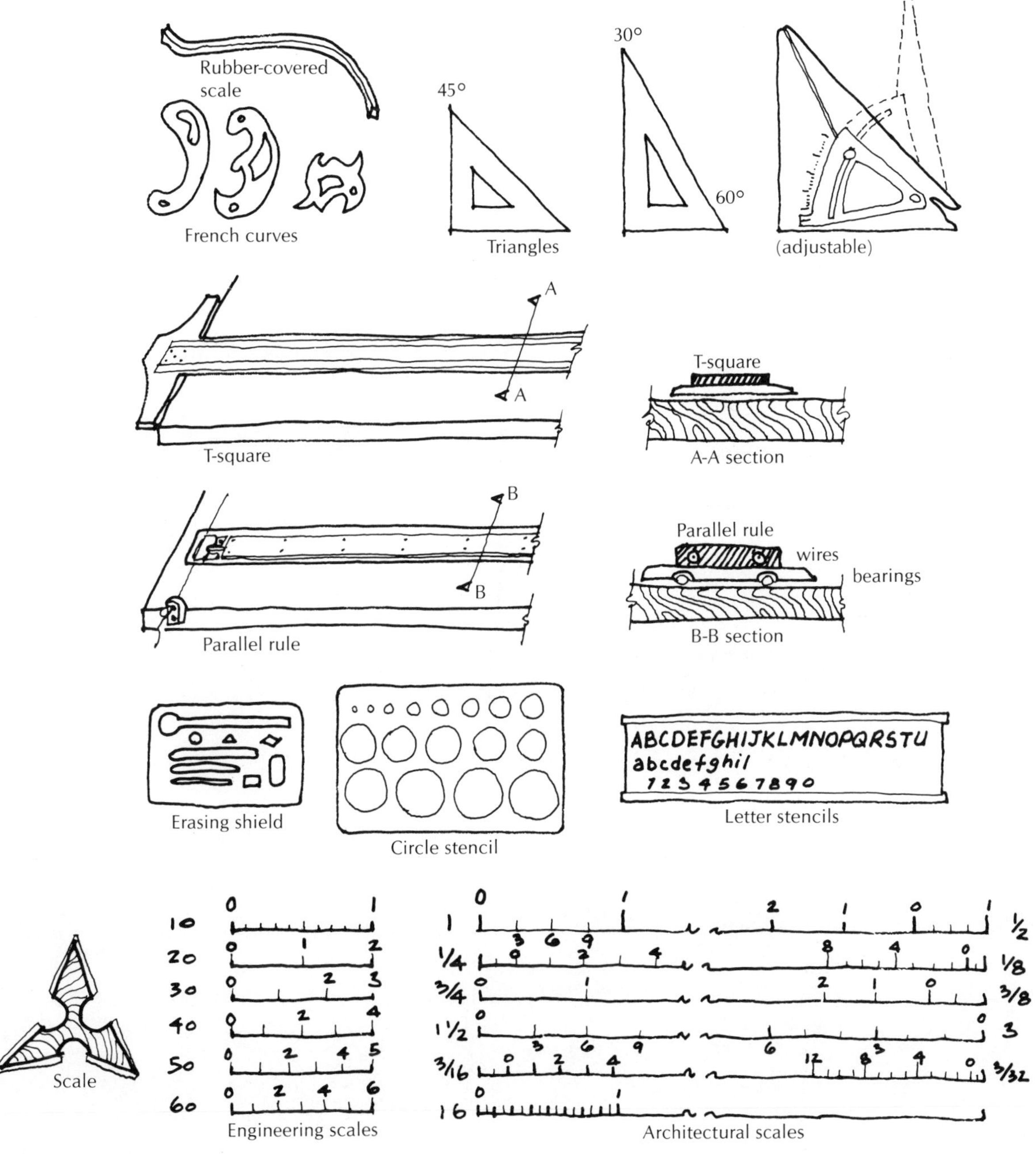

Figure 10.13 Drafting tools.

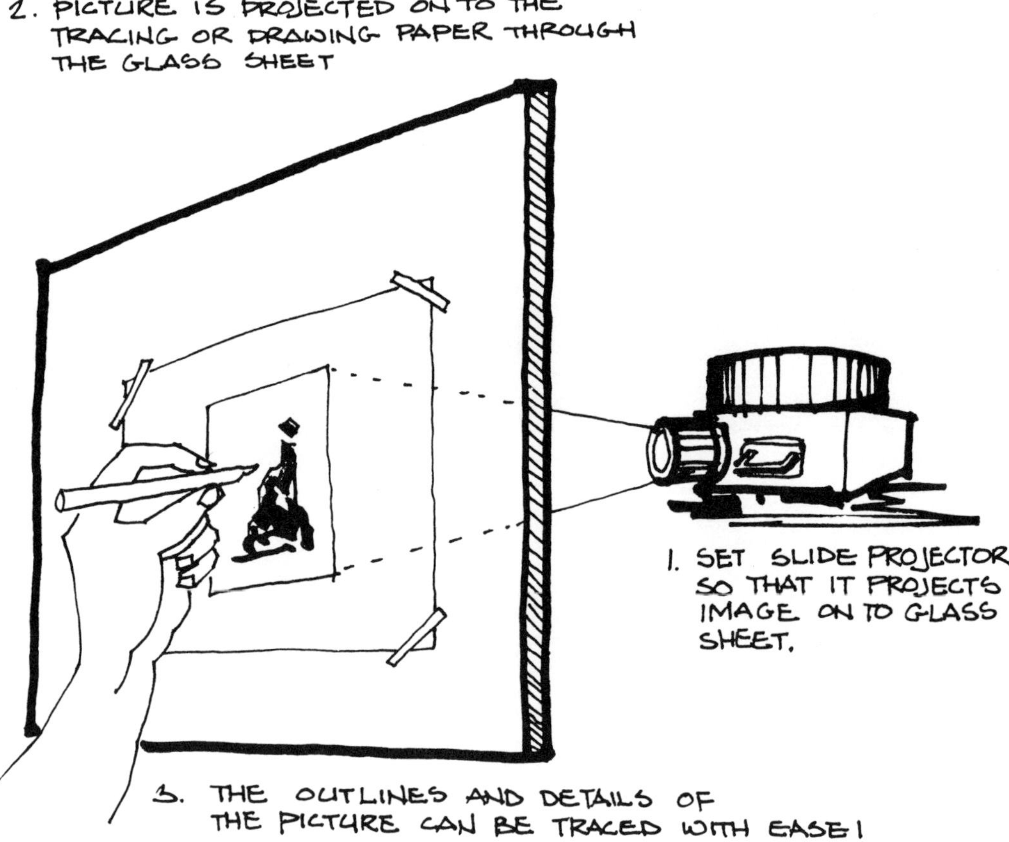
2. PICTURE IS PROJECTED ON TO THE TRACING OR DRAWING PAPER THROUGH THE GLASS SHEET
1. SET SLIDE PROJECTOR SO THAT IT PROJECTS IMAGE ON TO GLASS SHEET.
3. THE OUTLINES AND DETAILS OF THE PICTURE CAN BE TRACED WITH EASE!

Figure 10.14 Making drawings from slides.

black and white or instant-camera photograph on to an acetate sheet and project this by means of an overhead projector to enlarge the image to a desired size from which a final tracing can be made.

The projection method can allow one to make simple black-and-white drawing from slides that can then be reduced for use in reports. High-contrast photographic papers are used to create stylized images from black-and-white photographs. Conversely, slides may be made from a variety of secondary sources of information. Old documents, photographs in historical journals, title sheets of legal documents, excerpts from newspapers and journals, cartoons, and diagrams can be reproduced into photographs and slides. This material, if it is well integrated and woven through oral presentations or written reports, can add an extra stimulating dimension to the work. Accompanying music or taped recordings of sounds in places being visually depicted are other, similarly reinforcing techniques that can aid graphic presentations.

FORM AND STYLE

The form and style of graphic presentation to use in a particular situation should be appropriate to the task at hand. In a preliminary presentation, the graphics should reflect the tentative nature of the work. Initial rounds of dialogue with a client group may consist of very "soft" copy, of chalkboard sketches that can be modified and erased in response to the audience reaction and as part of the teaching-learning dialogue underway. Merely to label a presentation or a set of graphics preliminary or tentative is not sufficient. The form, style, and organization of the work must show openness and possible areas of change. Preliminaries should look like preliminaries, not like final contract drawings. I have worked in offices where I have made freehand tracing paper overlays of drafted, semi-developed drawings to use for preliminary presentations to the client. Clients such as citizen groups can become nervous, even hostile, if they are presented with what appears to be an accomplished fact when they were expecting an exchange of ideas, and understandably so. Similarly, final presentations should appear finished, organized, proofread, checked for accurate cross-indexing, and well integrated with the written material.

Graphics in Presentation

The importance of learning how to make an organized and convincing presentation cannot be overstated. More and more as planners assume more than a technical adviser role, they are in the business of selling their ideas about a professional problem. Graphics must be an integral part of both written and verbal presentations.

In oral presentations, the graphics conventionally used consist of slides that are projected or drawings mounted on chip boards and installed for display. These may be of people and places, or charts, maps, and diagrams. Graphics that can be mounted and hung on walls or display screens have an advantage over slides, for if they can be left in place, they can be examined by the audience before and after the presentation. Sometimes three-dimensional models of the site and project can have the same advantage. On the other hand, slides, including slides of maps and drawings, have the advantage that they can be presented one at a time, under the control of the person making the presentation, who can stress those aspects that he or she deems important. Showing slides and commenting on them thus gives the planner more control over what the audience pays attention to than installing graphic boards. Finally, reductions of some of the material presented or a "program" or "guide" of what is going to be presented may be included in a one- or two-page handout distributed to the audience.

Equipment like overhead, slide, or movie projectors or videotape and tape recordings should be checked to be sure they work and are positioned properly before a presentation. This is particularly critical if two or more projectors are being used and one is attempting to fade in and out of images by using special "dissolve" units on projectors. It is disconcerting to both the audience and the speakers if mechanical difficulties disrupt the flow of the presentation. If the audiovisual equipment is an important part of the presentation, the professional must make sure the right equipment is on hand and must try it out prior to use. Replacement items such as extra bulbs must be kept on hand. The films, tapes, and slides to be used must be run through at least once. When a slide sticks in the middle of the presentation it is usually not due to bad luck but to a bent slide, one that could have been weeded out if a trial run had been made.

The presentation should be kept within the limits of the time designated. If you overrun, you may not be allowed to finish your conclusion and recommendations. Attention spans are short, so try not to tax them. Large numbers of slides can put an audience to sleep, particularly after dinner.

One way to keep an audience's attention is to have a good mix of both the subject material you present and of media. For instance, graphics can be of physical spaces in the form of maps, plans, or drawings; of data in the form of charts, diagrams, and graphs; of people to lend some human interest and dynamics; and historical pieces that establish a continuity with past traditions. A good mix of graphic types—charts, slides, photographs, diagrams, and handouts—coupled with perhaps more than one person to make the presentation can liven the proceedings. Memorable presentations can consist of presenters role-playing a city planning commission hearing or a formal advisory panel, to the delight and amusement of the audience, as well as convincingly communicating their point of view. Sometimes lighthearted touches, like using flamboyant hats to signify changes in role being played or having the drollest in a team of professionals hold up charts and graphics, can win the audience's attention and convictions. This is not to suggest that such techniques should be tried in a presentation before a formal corporate board or important political gathering. The medium must fit the message and the audience. But a little loosening of the conventionally accepted professional style may go a long way toward gaining rapport with the audience.

Professionals using graphic illustrations in presentations before groups should stand beside the chart or diagram and talk out to the audience. It is very disconcerting to listen to a speaker address himself to the map on the wall. The size of the drawing, the blocks and colors used, the size of the lettering, and major symbols should be large enough to be recognizable at the back of the room. A good rule of thumb in deciding on letter size is to remember that a one-inch cap height type can be read by an audience at fifty feet, two-inch type at one-hundred feet, and so on, so that theoretically a six-inch letter can be read at three-hundred feet. Type that is reversed, white type on black, looks 10 percent larger than the normal black type on white ground. If representatives of cable television and radio, newspaper reporters, and photographers are present, graphics should be available in black and white, and there should be handout materials for dissemination. Providing handouts of well-organized material enhances one's chances of getting good and accurate coverage.

Lighting

When slides or films are shown or the overhead projector is used, some light should play on the speaker, even if the only way to do this is to flash on the light from the projector, which the speaker can stand in front of. Ideally, the speaker's face should be lighted from above (from a ceiling light) rather than from below.

Handouts

A handout that uses the same graphics in reduced form is useful. Graphics can provide the needed continuity between the various stages of a presentation when they are reproduced in different forms. For example, a handout for a verbal presentation can be modified to become the cover page or abstract page of a final report. Keeping the same format and layout can induce a sense of familiarity with the material that might help the reader's approach to the finished work. A handout should be a guide or a program to the presentation. The audience's attention should be drawn to it so they understand what it contains. Then the presenter should refer verbally to it throughout the talk, focusing the audience's attention on it. Otherwise the crowd will generally not read a handout, especially if the presentation is lively and informative. However, the verbal and visual message can be reinforced by repeating key words and figures in the handout. The professional's job is to see that the handout is read and may in fact create enough interest to get taken home and referred to.

Use of boards

Some thought should be given to the sequencing of diagrams and charts on boards at a presentation. They can be arranged in the order they will be referred to in descriptions so that in reviewing the presenter can just go down the line and summarize what has been said. Graphics should use only key words. Large-sized, bold-lettered headings will be retained better by an audience than will smaller-sized listings and itemizations. The graphics on the board should be grounded with strong base lines or a border so that they do not appear to float unconnected to each other or to the edges of the board itself.

Explaining graphics

If you are using a map juxtaposed with slides to "walk" people through an area, point out on the map the vantage point from which the slide is taken, and move back and forth between the map and the slide so the audience understands the connection between the two graphics being projected. Those who use maps, plans, and elevations routinely in their work tend to forget that many people are not as well trained to visualize in two dimensions as they are able to comprehend in three dimensions. The layperson's inability to read maps was demonstrated clearly to me when I was attempting to get an illiterate older man in a village I was studying to show me which part of the village he lived in and which social spaces he used the most. He kept claiming that the map I had drawn up with much care and labor over several months was wrong. Suddenly he turned the map around 180 degrees so that the map was now aligned with the way the village was spread out outside his house, turned to me, and said, "Now you have it right." I could only agree.

Graphics in Reports

In compiling the final report, the planner should look at all the graphic visual material collected and produced for the project up to that time and plan to integrate as much of it as possible in the final report. Using some of the material that has had favorable response during the intermediate stages of the project will, when used in the final report, give a sense of continuity to the reader who has been involved previously with the project.

Modifying some of the work slightly, reducing it in size, and adding labels or explanatory notes and legends to existing graphic work may make them appropriate for the final report. Graphics can be used to provide space breaks or breathers to lengthy technical or conceptually complex paragraphs. Spacing and size of the graphics should reflect the main purpose of using that graphic in that location. Headings, titles, and legends to the graphics should emphasize and clarify certain points; they should not obfuscate the point with professional jargon. Graphics should tell the viewer what he should see, in simple words. One word, a sentence, perhaps even a line from a poem will do. Generally a graphic with one message is more effective than one with several.

A dummy of the report can be used to plan blocks of space in which various bits of information, artwork, maps, and diagrams should go and to ensure a flow of the narrative and the accompanying illustrations. A dummy gives a sense of how long the report might become, what additional work must be completed, the costs of reproduction, and so on. This information can be used to schedule a team's work, decide on the quality and cost of reproduction techniques, use of colors, and other such questions. (See Figure 10.15.)

Accepted conventions should be adhered to, as far as possible, in the ways the material is composed. If color is used to indicate land use, green is the conventional color for vegetation, parks, and recreation, red for commercial, yellow for residential, and blue for industry. Certain standard graphic symbols are used in planning, and books are available that give good examples. Copying these and using the more conventional, stylized symbols on a drawing is preferable to realistic depictions that can appear amateurish when rendered by less artistic and neophyte planners. (See Figure 10.16.)

Certain graphic conventions are part of our daily awareness; for example, street signs generally use dotted lines as tentative suggestions (overtake with caution) and solid lines as definite directives. Try not to fly in the face of such conventions. Be aware of them in your daily life and reflect them in your work. For instance, in representing part of a project that is

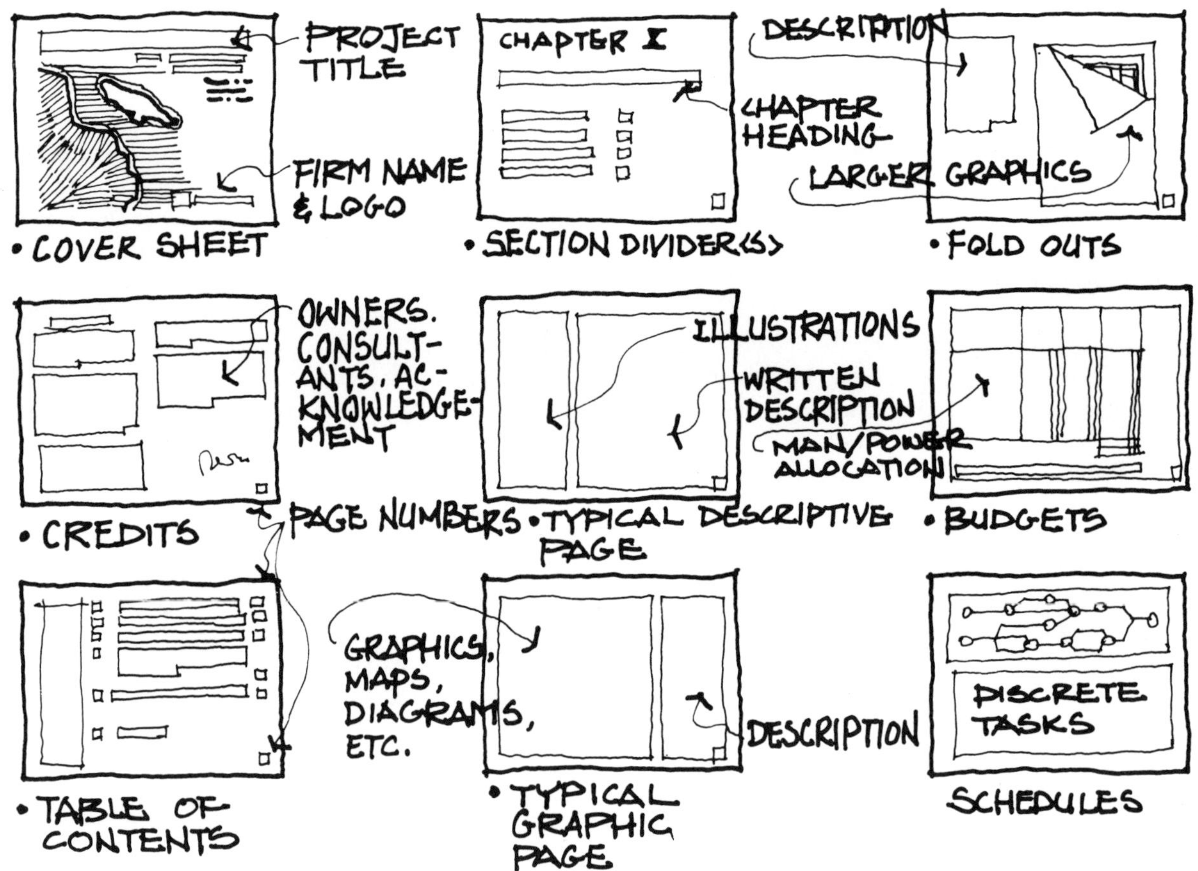

Figure 10.15 Preparing a dummy for a report.

complete and another part that is proposed, the completed portions can be indicated in bolder, more vivid colors than the proposed. The eye should see what is foreground and what is background, what is complete and final and what is tentative in a presentation.

A report should not be padded with repetitions of maps and graphics that are not enlightening or useful in further clarifications. To the reader who is paying attention, such padding is annoying; to the one who is skimming, it is confusing.

Planners must leave enough time when completing a final report to go over the finished work, to make sure all pages, including those containing figures, illustrations, and diagrams, are numbered, that cross-references are correct, and that the table of contents is complete. In the atmosphere of crisis that prevails when a project is being completed, leaving aside the final two or three hours for such checking is difficult, but a well-integrated, complete manuscript evokes confidence in the author and the work, and it is generally worthwhile to have one less illustration or one less paragraph in the report and have a thoroughly completed and cross-checked document. There is nothing so dismaying as to find, after one has made fifty copies of a manuscript, that the client's name is

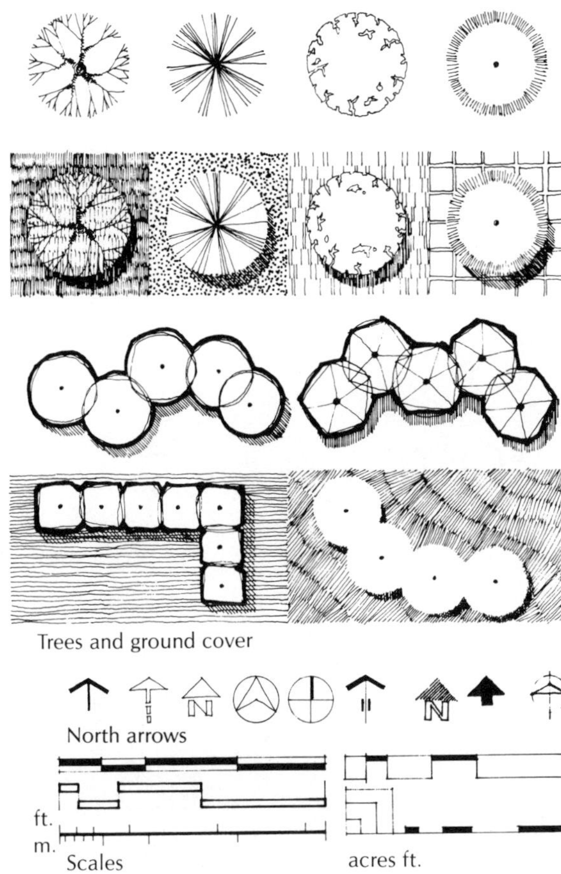

Figure 10.16 Stylized symbols.

missing or, worse yet, misspelled, that pages are out of order, that tables that are referred to are not included or do not show what they are supposed to. Such mistakes do not inspire confidence in the author's professionalism. Difficult though it may seem at the crisis hour, the time to check the report is before duplicating and disseminating it.

FOR FURTHER CONSIDERATION

1. Review a planning report from a local or regional planning agency. Examine the categories of graphics used and evaluate how well they are integrated into and augment the text. Note places where additional graphics would have made for more effective communications.
2. Critique a report you have worked on along the same parameters.
3. Put together a dummy for a report you are compiling. Break down the needed work into appropriate categories and estimate time and skills required and cost of reproduction. Keep accounts of actual time and materials expended in completion and compare with the original estimations. Identify what items you underestimated or overestimated.
4. If you had to compile another report similar to the one you have just completed, what aspects of the graphics would you change in both planning and execution?
5. Attend a public presentation of a planning project and evaluate the graphics used along the following criteria:
   a. Subject matter, its interest and evocativeness.
   b. Quality.
   c. Type.
   d. Integration with oral presentation.
   e. Efficiency and operation of audiovisual machinery.
   f. Balance between verbal and graphic components of the presentation.
6. Critique your next presentation using the criteria in number 5.
7. For this next presentation, prepare a handout for the audience and a media release for newspaper journalists and television reporters.

BIBLIOGRAPHY

Although a number of books cover graphic, drafting, and rendering skills for professions such as architecture, landscape architecture, and graphic design, there is no standard book on useful graphic techniques for planners. Planners therefore have to read about selective techniques in books aimed at these other related professions. The following books are useful in the areas described.

Babbie, Earl R., *Survey Research Methods,* Wadsworth, Belmont, California, 1973.

Chapter 13, 'Constructing and Understanding Tables," is especially useful.

Berryman, Gregg, *Notes on Graphic Design and Visual Communication,* William Kaufmann, 1979.

Contains useful sections on design, graphic design symbols, logos, pictographs, and type.

Ching, Frank, *Architectural Graphics,* Van Nostrand Reinhold Co., 1975.

A good preliminary section on basic equipment and materials, the site plan, landscape and ground textures, graphic presentation symbols, hand lettering, and architectural presentations.

D'Amelio, Joseph, *Perspective Drawing Handbook,* Tudor Publishing Co., New York, 1964.

Useful for those who have had no drawing courses to gain some understanding of perspective drawings.

Duncan, Robert, *Architectural Graphics and Communications,* Kendall/Hunt Publishing Company, Dubuque, Iowa, 1980, esp. pp. 1-21.

Useful illustrations and use of some basic drawing equipment.

Hartmann, Robert R., *Graphics for Designers: A Studio Workbook,* Iowa State University Press, Ames, Iowa, 1976.

Useful sections on the drawing process, graphic analysis, generating ideas graphically, line drawing techniques, felt-tip sketches, and photographic techniques and applications.

Kliment, Stephen A., *Creative Communications for a Successful Design Practice,* Watson-Guptill Publications, New York, 1977, esp. pp. 27-35, 61-73.

Although this book is aimed primarily at physical planners and designers, there are lots of cues in it to aid planners to organize and integrate graphics more effectively.

Laseau, Paul, *Graphic Thinking for Architects and Designers,* Van Nostrand Reinhold Co., New York, 1980.

Interesting sections on visual communications through time, graphic thinking, and abstraction. A general graphics book useful to planners.

Peña, William, with William Caudill and John Focke, *Problem Seeking: An Architectural Programming Primer,* Cahners Books International, Boston, 1977.

Index

About the Contributors

All of the contributors are affiliated with The University of Michigan, Ann Arbor.

Editor

Hemalata C. Dandekar, assistant professor of urban planning, teaches courses on planning techniques and urban and regional planning in developing countries. A graduate of the University of Bombay (B. Arch.), The University of Michigan (M. Arch.), and the University of California, Los Angeles (Ph.D., urban planning), Dr. Dandekar has also taught at the Massachusetts Institute of Technology. Her professional experience as architect-planner spans diverse regions and cultures and includes work in India, Japan, and the United States. She has worked as a consultant for UNESCO and the World Bank.

Dr. Dandekar is a Bunting Fellow at Radcliffe College. She is the author of several scholarly works including articles in *Ekistics* and *Economic and Political Weekly.* Her research interests include effective communications in planning and architecture and Third World development, particularly in the areas of appropriate technology, rural and urban connections, and women.

Contributors

Peter Ash, M.D., is on the faculty of the Department of Psychiatry. His interests include interpersonal processes in groups and families.

Allan G. Feldt is a professor of urban and regional planning. He has taught at Cornell University. Dr. Feldt's principal areas of expertise include: population estimation, projection and analysis; retirement communities; simulation/gaming; owner-built housing; decentralist planning; and energy conservation.

Nancy Nishikawa is a graduate of the Urban Planning Program and the School of Natural Resources. She is currently working on economic studies with the Washtenaw County Metropolitan Planning Commission and collaborating with Hemalata Dandekar on a book on women and development in the Third World.

Michell J. Rycus, after thirteen years as a scientist in the aerospace industry, became involved in planning in the early 1970s. He is an assistant professor of urban planning. His courses include analytic methods for planners, energy planning, and resource planning. His research efforts focus on the technical and analytic aspects of planning.

Rudolf B. Schmerl is an associate professor in the School of Education. He has lectured on proposal writing for the National Graduate University and other institutions, and was planning director at Wayne County Community College in that institution's initial year.

Alfred W. Storey is director of the University Extension and an associate professor of speech communication. He is a consultant to business, industrial, and educational groups and organizations on public speaking and interpersonal communication.

Katharine P. Warner, associate professor in the College of Architecture and Urban Planning, teaches courses in housing development and neighborhood planning. She is an active "public participant" member of Washtenaw County's Planning Commission and Ann Arbor's Public Housing Commission.